UFOs, Aliens and the Battle for the Truth

Pocket Essentials by Neil Nixon

Creative Writing

UFOs, Aliens and the Battle for the Truth

A SHORT HISTORY OF UFOLOGY

NEIL NIXON

Oldcastle Books

This edition published in 2020 by
Oldcastle Books,
Harpenden, UK
www.pocketessentials.com

First published in 2002 by Pocket Essentials

A CIP catalogue record for this book is available from the British Library.

ISBN
978-0-85730-431-5 (print)
978-0-85730-432-2 (ebook)

2 4 6 8 10 9 7 5 3 1

Typeset in 11.25 on 14pt Adobe Garamond Pro
by Avocet Typeset, Bideford, Devon, EX39 2BP
Printed and bound in Great Britain by Clays Ltd, Elcograf S.p.A.

Dedicated to two women without whom this book, and my life as I know it, would not exist.

My wife Jane whose understanding of my UFO obsession is one of the many loving and selfless acts that keep our love alive. I love you, Jane.

and

Jenny Randles, a persistent and dedicated researcher unafraid of the truth, and a worthy role model to anyone intent on making a contribution to UFO investigation.

CONTENTS

CONTENTS

Introduction

One of These Days We Might Be Smart Enough to Ask the Right Questions

We'll get to UFOs in a few seconds but first let's talk about the title of this book and how the book is set up. There is a 'battle for the truth' raging out there and there has been for most of the history of 'ufology'. So, this book is partly a primer on the main stories and the history of UFO sightings but it's also an account of the history of ideas, beliefs and the people who have made ufology what it is. In that context it makes sense to present a history of the whole subject and then examine key aspects (like the best cases) by re-presenting important bits of that history, so the unfolding of these stories and what we know to be true can be shown clearly. The subject is epic, and this book is far from epic in length, so the notes along the way and a helpful list of ten websites at the end could well be your starter for much more involvement. With that in mind the 'battle for the truth' here is often presented with a summary of differing opinions about what – if anything – explains significant cases in ufology. All this is designed to allow you to consider for yourself how credible the main ideas which attempt to make sense of the bigger picture actually are. If this book doesn't get you thinking as well as taking in some amazing stories then it's failed in one of its main aims. Speaking of which, this publisher and this series of books have a reputation justly built on succinct summaries of their subjects. This subject is different if only because so much of what we 'know' is debatable and so much of what ufology is about rests on the people who have made the subject. So, unusually for one of these books the author will intrude occasionally. It's a fitting way to guide you through the subject if only because I've been somewhere in this battle for the truth since I was at school. I'm still involved because I believe there is lots to learn and ufology remains one of the most amazing things anyone can study. I'm also still involved because – as I see it anyway – I've developed some survival skills along the way. As you'll see from what follows in this book, ufology can present a web of conspiracy culture and scientific conundrums. So, those survival skills are essential.

UFO means Unidentified Flying Object. The term was originally coined along with several others like UAO (Unidentified Ariel Object) and Flying Saucer to describe unknown objects seen in the sky.

The late and much missed ufologist Leonard Stringfield (1920-1994) once tried to capture the frustration of chasing the elusive evidence for crashed UFOs by titling a book chapter, 'The Search For Proof In A Squirrel's Cage'.[1] At this point in the twenty-first century, it is tempting to say that his estimate of the confusion and contradiction he experienced barely does justice to today's situation. It is a complex subject. The truth is out there somewhere and some of it may be in this book but, to complicate matters, you and I could easily find differing truths.

Celebrated ufologist Dr J Allen Hynek[2] noted that UFOs themselves are not studied. In reality, most of those with an active interest in UFOs only encounter verbal or written reports and reproduced images of the UFOs in question. This is true for both active investigators and armchair students of the subject. Put bluntly, ufology is largely the study of secondary sources of evidence. This situation frequently leaves doubt in the minds of some as to whether the objects in question were ever genuinely unidentified and/or flying. And it gets more confusing! The term UFO on a book or video jacket has proven a sales winner time and again but much of the most marketable material in the last few decades has not been primarily concerned with flying objects at all. Currently, UFO investigation and the popular market on the subject also include reports of other phenomena including: cattle mutilations; human abductions by aliens; people who claim to channel messages from aliens; and 'alternative archaeology', which presents a revisionist view of history in which alien intelligences play a pivotal role in the history of life on this planet – especially when it comes to explaining archaeological mysteries.

Central to all the strands of UFO investigation is that there is a series of phenomena that can be studied. Virtually all amateur UFO investigation assumes that there may be intelligence behind some of these phenomena. The most popular viewpoint amongst the subject's greatest supporters is that life alien to this planet is involved. There is certainly logic and rational thought behind these notions, but there is also much disagreement.

The truth about UFOs and UFO investigation is that a central core of mysterious reports are continually being appropriated and hijacked by people with their own agendas. The motivation behind this is often well-intentioned but the result has been to scatter the subject in a way that leaves entrenched camps seething with mutual suspicion and much research being undertaken in isolation. Information travels around, work is published and claims are made, but the major casualty is undisputed truth. The result is that many UFO cases of genuine substance are tainted by the shenanigans surrounding the investigations. Leonard Stringfield's 'squirrel's cage' comment pre-dated the mass use of the internet. You don't have to spend long online examining contrasting views to see how personal things can get amongst truth-seekers.

The many ironies are not lost on some of the key players. Years ago the British ufologist Andy Roberts unleashed the merciless and amusing newsletter 'The Armchair Ufologist'. His motto: 'Tough on Ufology, tough on the causes of Ufology.' This chronicle of political infighting and massive egos built on minuscule ideas casts ufology as a collision between support groups for the socially wretched and an exercise in self-aggrandisement for a select group of the terminally delusional. Roberts' agenda was, in fact, very positive. He and his colleague Dr David Clarke also wrote the sublime *Flying Saucerers*[3] – a social history of UFO investigation that reads like a plot outline for the greatest movie Terry Gilliam could ever make. Clarke also wrote a book entitled *How UFOs Conquered the World*. Not an account of an alien invasion but, according to its own subtitle, 'The History of a Modern Myth'.

Put simply, it is very often impossible to separate the claims made about UFO events from the people who make these claims. The vast majority of UFO case investigation is amateur and the vast majority of investigators undergo a rudimentary initiation at best. It is easy to condemn the chaos and comedy that often result but there is little or no alternative. In a quiet year there are a few hundred UFO reports in the UK and no professional organisation exists to monitor, investigate and report on the situation.

There is some professional investigation ongoing, some of it producing vital and challenging work. However, the bad press attached to UFO investigation has left the subject in an academic limbo. On the one hand, UFO reports are fascinating and more substantial in terms of evidence than the cynics would like to admit. On the other, many employed in universities and colleges regard their amateur colleagues in UFO groups as a kind of 'Care in the Community' branch of academia. The end result is predictable and tragic for the subject. An earlier edition of this book came out in 2002 and noted: 'UFO research has been seen as a certain route to career suicide for the best scientific and social scientific minds of several generations. Only a handful of serious, peer-reviewed studies exist. Research undertaken in psychology, tectonics and sociology has made a substantial contribution to the UFO debate but it often fails to impress those involved in gathering field reports in their local area.' All of that statement remains true. In some cases, those in research groups simply don't understand the academic research. In most cases, they get the gist of the ideas but, understandably, point out that it doesn't help them to explain anything to the terrified witness they've just interviewed. The most damning argument from the rank and file is also the most obvious. The academics who claim to study UFOs seldom do the local groundwork or meet the witnesses. Much academic work concerns itself with trying to replicate UFO events in laboratories. The academic fraternity for their part have often slammed the primitive and inaccurate investigative methods of the self-appointed research community of ufology.

Ufology, a loose term coined to include pretty much any investigation related to UFO reports, is not a science. This was eloquently stated in a 1979 paper. NASA scientist James Oberg won a prize offered by Cutty Sark Whisky with his paper 'The Failure of the "Science" of Ufology'.[4] Presenting himself as a benevolent sceptic, Oberg demolished the pretensions of the fledgling science with some substantial points. He saw ufology as a protest movement, or the result of effective myth-making. Almost 20 years later I followed up his report with a much longer paper.[5] I found many of Oberg's points still applied, although the situation had become more complicated.

Ufology may not be a science. In fact, it is no one thing. UFOs and the study of UFO events resemble, by turns, a protest movement, a branch of the entertainment industry, a collection of religious movements, a well-established scam for fleecing the public and a bizarre fringe profession employing a collection of visionaries and mavericks. The people problem (i.e. those within the UFO community who gravitate towards the answers they want regardless of the evidence) may complicate the whole picture but we have to keep one truth in mind: UFOs and UFO investigation remain alive and well because, despite the problems, people continue to see and experience things they can't begin to explain and when these stories become public knowledge many others are fascinated to the point that the need to know grips them and doesn't let go.

Most of the people on the receiving end of such experiences did not ask for them and do not seek any publicity. They are, by turns, fascinated, frightened, changed as personalities and physically harmed. UFO cases may peak at times when UFOs are a popular media subject but they never go away. Many cases are easily explained. Some of the best known are kept alive by a mixture of faith and bullshit but there are others so indisputably mysterious that they present a challenge to established knowledge in a number of fields. The stories in this book include the terrifying, tragic and incredible. Honest, sincere people report such things every day. Some of the possibilities presented by these cases are so awesome they strike at the very core of the deepest questions we can comprehend.

Your Guide

Hello, I'm Neil and I've been doing UFOs all my life. Having stated that people with an agenda hijack the information, I had better be honest about myself. Then I'll mainly get out of the way and let the evidence do the talking.

The limitless imaginative possibilities of the stories got me involved in the field. Like most of my fellow ufologists I had it sussed within days. People were seeing alien spacecraft and anyone who disagreed was narrow-minded. Sorted! At this point I was around seven years old. Not long after that, I read the famous report of George Adamski's meeting with a man from Venus and thought, 'Whaattt!! Maybe some of these stories aren't true.'

Since then I've remained true to the UFO cause and the pursuit of critical thinking. I've been part of investigation groups, stood on the stage at conferences and waded through a library of books even when their covers alone convinced me that the most reliable information inside was the publisher's name and address. I'm still here because this subject, in my opinion, remains one of the greatest mysteries of the modern age.

I am sceptical about most claims and I'm highly sceptical of evidence gathered using hypnotic regression. Long ago I came to believe that some figures on the fringes at UFO conferences and media coverage, like Albert Budden, were making a massive contribution to our knowledge on the subject. I value the hard-headed scepticism of people like Brian Dunning[6]. But I also think that a few cases are so baffling in the face of all our current understanding that we can't begin to explain them. These are some of my favourite cases, along with the really strange and surreal ones.

I've had my own experiences, but I think them all explicable.

By far the most memorable was the time that government agents chased my car up a Kent road with the intention of apprehending me. That day taught me a lot. The story is coming up later in the book.

Right now, it is time for a brief history of bafflement.

Historical Precedents

The modern era of UFO investigation and popular appeal started in 1947 but UFO events are almost as old as recorded time. It is an article of faith amongst many UFO proponents that the ancient texts of several religions describe UFO events. This view is seldom shared by scholars of the religions in question. The claims of the UFO proponents revolve around the description of incredible events and their similarity to modern UFO reports. A further claim amongst the UFO community is that the events described from a range of ancient texts describe the same phenomena. The conclusion drawn is that this consistency is clear evidence that involvement with UFOs and their occupants was a regular feature of the lives of ancient peoples.

Erich Von Daniken's multimillion-selling *Chariots of the Gods*[7] and a host of other successful titles have presented a world in which alien spacecraft regularly landed in ancient times. Within the UFO community itself opinions are divided as to the meaning, if any, of these events. The most ambitious claims posit a kind of galactic garden theory in which aliens engineered Earth's biological and social future. These claims include the belief amongst some that aliens invented our religions and created mankind. This is often linked to the assertion in the book of Genesis that God made Man in his own image. The 'God' in question being an alien scientist. A variant on this story also supports religious beliefs in some UFO-related faiths, notably Raëlism which teaches that the human race as we know it was created scientifically by human scientists from space.

Such arguments have many detractors. Some say that the term 'alternative archaeologist' is a fitting description for some of the field's best-known proponents since their main archaeological tools appear to be an armchair, the internet, a pile of books and an active imagination. Alternative archaeology/ancient astronaut investigators, like many proponents of UFO existence, are derided for their reliance on sloppy and unscientific reasoning, but the criticism of the sceptics is probably most cutting for historical investigators whose evidence relies on interpretation of historic events and revisions of scientific viewpoints. Whatever the difficulties of sorting the sense from the nonsense, it's beyond argument that this variant of the UFO question is highly engaging and television series like *Ancient Aliens* have proven perennially popular.

In his considered demolition of ufology as a branch of science, James Oberg pointed out that the 'residue fallacy' in which some unexplained data are used to support a claim is plain bad science. This is certainly true of the basics of the ancient astronaut belief. Essentially, Oberg is saying that the evidence of one hard-to-explain event, like the destruction of the cities of Sodom and Gomorrah and the changing of Lot's wife into a pillar of salt, doesn't prove that aliens used atomic weaponry on our ancestors.

There are some gripping mysteries in ancient texts. The pillar in the sky that led the Israelites from Egypt in Exodus Chapter 13 is one prime case. The Israelites were familiar with comets and stars, so the description of an airborne fire at night and airborne pillar of cloud by day may suggest some unique phenomena.

The descriptions of many key events in ancient texts are vivid and spectacular. It is possible to dismiss many UFO-related claims about this material as supposition but some substantial investigations retain their power. NASA scientist Josef F Blumrich laughed his way through most of *Chariots of the Gods* and set out to mount a scientific assault on the one section of the book that presented serious evidence he could challenge – the section dealing with the experience of the prophet Ezekiel. Here Blumrich found some details he could develop. At the time Blumrich was chief of NASA's Advanced Structural Development Branch with a CV that included work on Saturn V rockets, satellites, Skylab and the Space Shuttle. He would go on to earn NASA's Exceptional Service Medal.

Working from Ezekiel's descriptions Blumrich discovered that he could develop technical drawings and specifications that resembled workable spacecraft. Having set out on a part-time quest to rubbish what he saw as fanciful claims, Blumrich, like J Allen Hynek before him, found his scepticism overturned by the evidence. As he noted in the opening of *The Spaceships of Ezekiel*[8], 'Hardly ever was a total defeat so rewarding, so fascinating, and so delightful!' Ezekiel's account is a riot of events: 'A great whirlwind came out of the North... and a fire unfolding itself... out of the midst thereof came the likeness of four living creatures.' However, it was Ezekiel's description of airborne wheels within wheels that prompted Blumrich's work.

Ezekiel described a wheel and Blumrich analysed the performance characteristics and appearance in the description. It led to scale drawings, a model craft that performed superbly in a wind tunnel and equations that suggested that all parts of the package, from the likely rate of burn of the propellant to the additional weight of the requisite radiation shield, were constructed to specifications the leading space engineers of Blumrich's era would recognise.

Blumrich was not alone in providing scientific support for alleged UFO events in ancient texts. Soviet physicist Professor Mates Mendelevich Agrest was one of a number of scientists to have outlined the obvious comparisons between the destruction of Sodom and Gomorrah and an atomic explosion. In his view, Lot's wife was too close to the event and, far from salt, she was probably reduced to a well-incinerated pile of ashes. Agrest – who died in 2005 – first produced a serious academic paper on aspects of paleocontacts (the theory that Earth has been visited by extra-terrestrials) in 1959 and was a major influence on the beliefs and claims that followed.

Astrophysicist Carl Sagan[9] had a high-profile career that included cutting-edge work for NASA and an abiding interest in the search for extraterrestrial life. Sagan had little truck with much popular ufology, frequently acting as an erudite sceptic, but in 1966 he co-authored *Intelligent Life in the Universe*, a book which openly speculated about extraterrestrial visitations to Earth in the distant past. This book and a number of others brought about a golden age of ancient astronaut writing in the 1960s and 1970s.[10] The quality of writing and research in the area, both then and now, is variable. Ultimately, it is the evidence that keeps this aspect of UFO interest alive. Some evidence, such as the Biblical account that spawned Josef Blumrich's book, is really a matter of faith. There is logic in the attempts to revisit fantastic historic accounts in the light of current technology but we can't prove that Ezekiel encountered space travellers any more than we can prove the Resurrection to an atheist.

Another strand of evidence is more tangible to modern science. A series of artefacts including the pyramids and ancient maps has continually fed the belief in extraterrestrial intervention in our history. The basis of the belief is that the production of such artefacts was beyond the technology available to ancient peoples. An enticing mystery surrounds a series of ancient maps. The best known is Piri Re'is's portolan. A portolan is a practical map showing the route from port to port. Erich Von Daniken shows a projection of this ancient map and discusses its uncanny accuracy when plotted against a US Navy map from the 1960s. According to some analysts the staggering feature of the map is its detailed depiction of the east coast of South America and the coast of Antarctica. Piri Re'is was a Turkish admiral in the days when such employment generally amounted to little more than licensed piracy. The map was a distillation of a collection of earlier maps, believed to have been in the possession of Christopher Columbus. Re'is's professional belligerence finally caught up with him. His final public appearance was a crowd-pleasing gorefest involving the forcible separation of his head from his body in 1554. Officially, the coast of Antarctica was discovered in 1818. In reality, debate has raged for years over when some areas of land were discovered (as in mapped by western nations; clearly, existing inhabitants of these lands had no need to discover them) and the support for Piri Re'is comes from figures like scholar and civil engineer Arlington H Mallery who was a leading figure in arguments suggesting Columbus wasn't the first to establish contact across the Atlantic Ocean.

Von Daniken wasn't alone in finding significance in this map. Donald Keyhoe, a retired US Marine Corps major, detailed a discussion with Captain John Brent, a military training school buddy. In *Flying Saucers—Top Secret* (1960), Keyhoe quotes Brent as saying that the Piri Re'is map, 'is so accurate only one thing could explain it… a worldwide aerial survey.' This survey would have to pre-date the last ice age. The bold 'only one thing could explain it' assertion typifies the style of many of the claims in ancient astronaut/alternative archaeology writing. It is also demonstrably wrong. Professor of Anthropology Charles Hapgood was also aware of Re'is's map and many other intriguing portolans. His own research, outlined in *Earth's Shifting Crust* (1959) and *Maps of the Ancient Sea Kings* (1966), suggested the possibility that rapid shifts in the Earth's crust had moved continents thousands of miles, wiping out much evidence of civilisation and leaving a few fragments, like the accurate maps, with the survivors. Hapgood's theory fits wider legends of lost civilisations as well as most of the ancient astronaut beliefs. Ultimately, much of the evidence for ancient astronauts is a puzzling jumble that makes fleeting sense when woven into a legend that allows the brilliant skill of super-intelligent aliens to fill in the gaps. These stories may be true but, to date, all we've proven beyond doubt is the existence of the maps and the mysteries.

One of the most intriguing stories is covered in *The Sirius Mystery*, a work first published in 1976.[11] Oddly, the author Robert Temple was concerned about his bestselling book being considered in the same breath as works like *Chariots of the Gods*. By 1976 Von Daniken's work was already drawing unfavourable comments about the credibility of its claims. So, publicity supporting Robert Temple stressed the differences between *The Sirius Mystery* and *Chariots of the Gods*. Temple's work concentrates on the Dogon tribe of Mali and a belief in their culture that fish-like beings, called the Nommos, had visited Earth in the past. Temple's assured study links folk beliefs in both Africa and the Middle East, and establishes legends amongst the Dogon, including detailed knowledge of astronomy and physics, which appear to have beaten modern science to key discoveries. Notably, the knowledge that the star Sirius has a heavy companion orbiting the main star once every 50 years. If, as many assert, this is folk belief handed down over thousands of years, then it predates the discovery of Sirius B and the subsequent discovery that it was a super-heavy White Dwarf star. The accuracy of much Dogon knowledge is beyond question and the difference between their own folk legends and those of most other African tribes also lends weight to this ancient astronaut theory. Sceptical views – including those of James Oberg – have suggested that Europeans taught the tribe. Carl Sagan's book *Broca's Brain* (1979) points out that the Dogon were unaware of a ringed planet beyond Saturn in our own solar system (a significant observation since it suggested that the Dogon had the same knowledge gaps as other astronomers on Earth). The Dogon tribe are convinced that there is a third star, Sirius C. If the star or its remains are ever discovered, remember where you read it first.

These stories are a representative handful in an area of investigation that posits a range of alternative cosmologies and histories of our planet. Much of the alternative archaeology movement concerns itself with reinterpreting past societies and a good deal of this work owes little to present-day UFO investigations and ideas. At the other extreme there are ideas and investigations of mind-numbing complexity. Ancient astronaut theories have some strange relatives. Some investigators firmly believe that alien visitations are the result of highly evolved beings from our own history maintaining their contact with Earth. The survivors of Atlantis are a predictable addition to this debate but they are not alone. British ufologist David Barclay is one of a number of researchers who have argued that dinosaurs evolved into the well-known grey aliens of the present.[12]

The publishing explosion that followed *Chariots of the Gods* continues but, as with most areas of the UFO debate, the internet is now the real home of unfettered coverage of the UFO mysteries of the ancient past, along with niche-interest television channels. If this opening section has you itching to know more there's a supply of episodes of *Ancient Aliens* sufficient to binge-watch yourself into a stupor. If you want the shortest route to an erudite demolition of most of the major claims Wikipedia (an institution we'll discuss later in this book) remains resolutely sceptical and there's an explanation of the strange phenomenon known as the 'Gish Gallop' in a *Smithsonian Magazine* blog that is your starter for ten for spotting those moments when ancient astronaut theory might spill over into complete confusion.[13]

Fairies, Flaps and Fundamentals

Recorded history over the last 2,000 years shows us how everything we know as normal, everyday life evolved. It also suggests that UFO events, in all their varied forms, have been a constant throughout this time.

We have a strong historical record for most of the last two millennia. This record includes details of religious beliefs, folk wisdom and social panics – all of them relevant to the way people understood their strangest experiences. At times, these records make it clear that people saw strange things in the sky or met strange beings, and struggled to make sense of what was happening. There are thousands of sightings recorded over the last few thousand years: lights in the sky; abduction experiences; the utterly incomprehensible; and the occasional hoax. Making sense of the stories at this distance and with no physical evidence is impossible. Using them to inform our current understanding can be useful. The past lacks photographic and radar evidence but that is about the only significant difference.

There is no shortage of images. Pictures from China and Persia show airborne craft looking for all the world like small boxes with people inside them. Strange astral events find their way into a series of pictures, often showing unfortunate earthlings throwing back their heads and gesturing at the sky. Probably the most famous image of this time records the appearance of black and white globes over Basle in Switzerland in 1566. It shows a sky crowded with ufology's answer to Piccadilly Circus whilst a stunned group of onlookers show that overacting was a problem long before cheap Hollywood B-movies. Elsewhere, the technology of the day clearly affected the way people represented the reports. An etching currently in the possession of the New York Public Library shows aerial ships over France. We are talking ships with sails and rigging, observed by the predictably stunned onlookers on the ground. One witness in a small rowing boat is also, unwisely, standing and reacting theatrically.

Then as now, people read significance into such events. On a number of occasions royal proclamations or parliamentary time somewhere around the world resulted from UFO reports. It has been spuriously reported that the Emperor Charlemagne once outlawed aerial travel. He did, however, decree 'there shall be neither prognosticators and spell-casters, nor weather-magicians or spell-binders, and that wherever they are found they either be reformed or condemned'; a timely intervention when belief in the paranormal stretched to the point some considered it possible that beings living in the atmosphere controlled the worst weather events.[14]

Distant historical accounts contain their rarities. Octagons have never been common in UFO reports but one such object was observed on 15 April 1752 near Stavanger in Norway. They also show a gradual move toward the patterns of today. By the end of the nineteenth century UFO flaps were also becoming common. 'Flap' is another loose ufological term, generally used to describe a profusion of events that centre themselves at a particular time and place. There is no agreed number of reports or any definite time period applied to a flap.

A rash of reports from Wales at the start of the last century represents an early British flap. A brilliant disc over the principality on 1 February 1905 was followed by an intensely dark object over Llangollen on 2 September. On 18 May 1909, a man from Cardiff, walking near Caerphilly, reported an event that would find close parallels with many later reports. He encountered a large cylindrical object next to a road. Two odd-looking men in fur coats were inside, clearly speaking some strange foreign language. As the man approached, the object took off and headed away at great speed.

This final report bears some similarity to a rash of 'phantom airship' sightings that plagued the USA between 1896 and 1897. At the time it was widely believed that at least one brilliant inventor was at work on a craft that would revolutionise air transport. However, there is virtually no chance that such a vehicle was remotely close to a prototype. The vivid reports that became a regular feature of the papers pushed new boundaries in detail and helped to establish several staple items in UFO reports. In one case a railroad conductor, James Hooten, claimed to have come upon a landed airship. He described 'wheels' with blades driven by jets of air blown onto them. The captain of the ship had a brief conversation with Hooten during which he explained that the ship used 'compressed air and aeroplanes'. If this sounds familiar then remember that the bestselling fiction of Jules Verne was written around this time.

The airship legends continue as a substantial part of most detailed histories of the subject although it is almost certain that the major stories that drive this area of accounts are hoaxes. An animal abduction and mutilation story in which a rancher was said to have found an airship hovering over his cattle, before plucking one on a thin line and leaving it decimated, was concocted by the rancher Alexander Hamilton. The most famous case of the whole flap involved a crash retrieval, including the body of the pilot at Aurora, Texas, on 17 April 1897. The *Dallas Morning News* report stated of the pilot: 'while his remains are badly disfigured, enough of the original has been picked up to show that he was not an inhabitant of this world.' The ship also, apparently, contained writing in 'unknown hieroglyphics'.

Since the flap of 1896/7 the stories have circulated widely, appearing in various guises in the UFO literature and being discussed in terms of their importance to the modern day. It is beyond dispute that we have templates we would easily recognise in modern stories. James Hooten's encounter is typical of many, before and after, in which some aspect of technology just ahead of that known to humanity at the time was confidently demonstrated by a UFO occupant. Cattle mutilations continue to be reported to this day, many showing a clear resemblance to Alexander Hamilton's hoaxed claim. In other famous crash retrieval stories we find reports of strange hieroglyphic writing on debris from the wrecked UFO.

Extensive writing has been done on the US airship flap but the few who have checked the original sources have found clear evidence of hoaxing. Alexander Hamilton was a member of a Liars' Club, of which many existed in rural parts of the USA at the time. There seems little evidence that anyone around Aurora, Texas, ever took the crash retrieval seriously. Recently the story has represented a local nuisance as ufologists push to open graves and exhume the remains of the dead pilot. There is a succinct dismissal of the whole hysterical episode as the opening chapter of a co-authored work *The UFOs That Never Were*[15]. The odd aspect from this distance is the way that admitted hoaxes from the late nineteenth century bear an uncanny resemblance to apparently sincere reports in the present day.

Nowhere is this more marked than in the overlap between reports of meetings with creatures from alternative realities, like fairies and nature spirits, and present-day abduction reports. Substantial histories of such encounters have been produced, notably Walter Evans-Wentz's *The Fairy-Faith in Celtic Countries* (1911) and Katharine Briggs's *The Fairies in Tradition and Literature* (1967), both of which chronicle complex beliefs, multiple eyewitness accounts and stories of encounters between strange beings and humans which resemble the strangeness and intricacy of present-day abduction reports. This whole aspect of UFO investigation has produced a copious and contradictory literature. Crudely, researchers chase two lines of argument. A literal approach claims that reports of encounters with goblins, fairies and their ilk are the results of meetings between our simple-minded ancestors and aliens. A more involved argument forms part of the so-called 'new ufology'. This is a branch of investigation taking a broad view and allowing ideas from mysticism, folklore and psychology. These ideas are used to look at UFO events as potential evidence of a range of phenomena including the convergence of one or more dimensions of existence. Less literal than the nuts-and-bolts investigations, new ufology can claim one of two fathers: Carl Jung or Jacques Vallée. Jung's *Flying Saucers: A Modern Myth of*

Things Seen in the Sky (1959) was years ahead of its time and the first serious book-length study to examine the subject from within the minds of experiencers and those who believed the tales. In the preface to the English edition he states plainly that his work asks 'Why should it be more desirable for saucers to exist than not?' Vallée's early works[16] broke new ground in opening up investigations of present-day UFO cases in conjunction with established traditions of folk belief. His most influential work is arguably *Passport to Magonia* (1969 France, 1970 UK) which includes lengthy discussions highlighting the common ground in historic stories of encounters with beings from the fairy realms with more recent tales of encounters with aliens. Difficult to categorise but often revered by those most sympathetic to the new ufology are the works of John Keel (1930-2009). His book *The Mothman Prophecies* (1975 and revised subsequently) is an insider account of varied paranormal goings-on, mainly in Point Pleasant, West Virginia, in which the baffling events appear to have a purpose. Keel is caught in the middle of experiencing and attempting to interpret and understand, and his reports from the front line include astute observations such as 'once you have established a belief, the phenomenon adjusts its manifestations to support that belief and thereby escalate it'. Keel's tale ends, ultimately, in a tragedy for the town. The whole complex story eventually found its way into a Hollywood movie in which Richard Gere played Keel.

It is easy to overlook the fact that for most of the last 5,000 years so-called superstitious beliefs have been part of everyday reality for many people around the world. It is only in the recent past that they have been relegated by rationality to the fringes of our lives. This relegation has, in any case, struggled to succeed. Belief in the power of astrology, the reality of angels and the presence of spiritual entities within nature remains stronger than many people realise – though a few minutes surfing sites on these subjects would soon show you the diversity and passion behind current beliefs. The lore surrounding such beliefs has built up over centuries. Many themes which would subsequently become staples of UFO cases consistently appear in tales of the fairy kingdom. Many attempts to remove items from the fairy kingdom as proof of a visit appear to come to grief. The beings from the kingdom seem able to shape-shift and often appear in unfamiliar and confusing disguises. Time spent in the realm of these elemental beings may pass at a different rate to time for those left behind. In one case from Wales a man missing for three weeks believed he had only been gone for three hours. Messages gathered from the fairy kingdom may be contradictory and predictions have a tendency to amount to nothing. However, despite these problems, there is a peculiar consistency. The people contacted are often left clinging to stories which do them no favours in terms of their credibility. One of my favourite UFO cases of all time started its high-profile life in the writing of Jaques Vallée. Many of the nuts-and-bolts UFO crowd don't rate the bizarre case of Joe Simonton and his pancakes at all.[17, 18]

Simonton, a chicken farmer from Eagle River, Wisconsin, spent a bizarre few minutes on 18 April 1961 doing the bidding of a trio of small men inside a shiny craft. He gave them water; they gave him some rough and tasteless pancakes, one of which was later analysed by the Food and Drug Laboratory of the US Department of Health, Education and Welfare. The analysis, carried out at the behest of the US Air Force, found the pancakes to be quite ordinary and 'of terrestrial origin'. Support for Simonton came from J Allen Hynek who was called to investigate that case and concluded, 'There is no question that Mr Simonton felt that his contact had been a real experience' – though Hynek considered a waking dream state to be the likely explanation. Simonton's case corresponds with reported fairy lore in which food loses its flavour and all sense of life when passed from the fairy realm to our world. His sincere account of an experience that belongs in a third-rate sci-fi novel is also another regular feature of fairy lore.

Had Simonton's encounter happened a century earlier he would doubtless be regarded as, quite literally, telling a fairy story. His case belongs instead in the 'High Strangeness' files of the modern UFO era, an era we must now visit. This section has demonstrated that much of what we will find there may have roots that stretch back centuries. The present age, the age when 'Flying Saucer' and 'UFO' became dictionary entries and labels likely to generate income, started on 24 June 1947.

Okay Ken, Open Those Floodgates

On that date Kenneth Arnold was flying his CallAir aircraft near Mount Rainier in Washington State. He joined the search for a missing C-46 transport plane. He didn't find it. The wreckage and bodies would turn up soon enough and make a fleeting news item. Arnold, by contrast, would be credited with starting a phenomenon.

Arnold sighted nine objects flying in formation. Startled at the way the sun glinted off them and their rapid movement he began to take crude measurements. He estimated their speed at between 1,300 and 1,700 miles an hour, easily in excess of the fastest aircraft of the time. Using another aircraft in the sky, a DC-4 at some distance, Arnold managed to make some estimates of the size of the objects and length of the formation. He believed the objects to be two-thirds the size of the DC-4 and estimated the formation to be spread over five miles of sky. Arnold's initial reaction was to believe the objects were military hardware, possibly missiles, under test. He radioed his sighting ahead and discussed what he had seen with fellow pilots after landing at Pendleton, Oregon. Within hours the news wires were starting to buzz with the story that heralded the era of the flying saucer.

Arnold's sighting is unremarkable in the context of the mind-numbing conspiracy claims that stalk the current literature. His conviction regarding guided missiles came from his initial discussions with other pilots. He would soon be persuaded towards an alien interpretation. The explosion of publicity that greeted the case unleashed a hysteria that would resolutely refuse to calm down. In retrospect, Kenneth Arnold's sighting is the perfect case to open the era of UFOs as a popular interest. Apart from anything else, the case has something for everyone: a vivid image on which to hang the whole story; odd twists; incredible speculation; and an unsung hero.

The claims made by Arnold were repeated around the world. Arnold became a celebrity and remained a figure of some veneration in the UFO world until his death in 1984. The dissection of the case by the media soon established that nothing man-made could achieve the reported speed of Arnold's objects. Within weeks further reports were coming in. Arnold himself fielded a fusillade of phone calls and letters, many reporting similar sightings and a substantial proportion firmly in the 'Thank God, at last somebody understands me' camp. The image of the little plane sharing the sky with a saucer squadron was sensational and allowed room for public speculation. There was no obvious conventional explanation. Given the scale of other reports and the lack of any obvious new military hardware matching the incredible performance of Arnold's flying saucers, the common consent was that Kenneth Arnold had seen alien spacecraft.

Nothing pushed the cause of this case more than the term 'flying saucers'. The term and the consequent explosion of saucer popularity provide insights into the way truth and apparent truth develop in UFO cases. Kenneth Arnold described crescent- or boomerang-shaped objects. The term saucer came from his comment that the objects moved 'like a saucer would if you skipped it across water'. Bill Bequette, a reporter for the *East Oregonian*, heard the description and coined the term 'flying saucer'. Whatever the accuracy of his report, Bequette remains one of the UFO world's unsung heroes. Without his crowd-pleasing summation of the incident we would have been denied a phrase that would popularise the whole business of sighting unusual airborne objects.

Nobody was with Arnold in his plane but interpretations of his experience had already distorted the truth. Within days the popular image of flying saucers and discs became the norm in other reports. The first instance of the word saucer being used in relation to UFO cases goes back to the American airship flap of the previous century and, as it had exactly 50 years before, the news industry once more made money on the back of UFO reports. Arnold's efforts to measure speed, size and distance gave his report an air of authority. But nobody else was there and his credibility has been called into question as succeeding generations of investigators have sought to put their own spin on the story. Alien spacecraft remains the popular option to explain what he saw, especially in the stream of big-selling books that take the angle that aliens are regular visitors and/or in league with at least one government on Earth.

Several commentators have highlighted the contradiction between Arnold's report of crescent or boomerang-shaped objects and the later popular acceptance of saucer-shaped craft. A photograph exists of him holding a drawing showing the shape of the objects. Arnold would go on to write an article stating 'I did see the discs' and a book called *The Coming of the Saucers* (1952). Amongst others, David Barclay goes as far as to suggest that Arnold was a covert intelligence operative involved in trying to bring the whole idea of UFOs into ridicule.[19] By comparison, James Easton's detailed analysis in *Fortean Times* gathers the information into a well-referenced argument suggesting that Arnold simply made a mistake when spotting a flight of white pelicans.[20] Predictably this claim went down like a lead brick in cyberspace and public houses frequented by people who believe firmly in an alien presence on this planet. At first glance Easton's claim appears to be the ufological equivalent of discovering a lost handwritten note in Marx's original manuscript for *Das Kapital* reading, 'This garbage is gonna make me rich!' It disturbed some of the alien conspiracy crowd all the more because, on close inspection, Easton's analysis is logical, researched to exemplary standards and supported by solid evidence. It doesn't rate high in the crowd-pleasing stakes, but it may well explain the events of 24 June 1947.

Easton's argument is specific to the Arnold case but we should note that speculation on life in outer space and the potential for alien visits was well advanced in 1947. In one notable event Dr Lyman Spitzer Jr, associate professor of astrophysics at Yale University, speculated on radio that Martians may have already visited Earth. The day after the programme a national US newspaper ran a report on his ideas. The dates of the radio show and the newspaper report are 22 and 23 June 1947, the two days before Kenneth Arnold's sighting![21]

But, just when it looks like a climate of mass awareness of possible alien life could have caused hysteria in Arnold and his audience, this case, like so many others, throws us a curve. A prospector on the ground at the time reported a sighting that appeared to support Arnold's report. The prospector's sighting lasted around a minute, allowing him time to use a telescope and focus on one 'disc' in a formation of similar objects. It is recorded in an FBI memo.[22] This confuses the issue but we can be sure that the popular image of Arnold encountering classic flying saucer-shaped discs is a myth. Another certainty is that the case made Arnold a more reflective and cynical human being. 'Believe me,' he said, 'If I ever see again a phenomenon of that sort in the sky, even if it's a one-story building, I won't say a word about it!'[23]

So, the modern era had begun. Since 1947 airborne objects have been a mainstay of UFO investigation. Such reports these days draw little excitement from some seasoned investigators. In the wake of Kenneth Arnold there have been spectacular cases, incredible claims and numerous precedents firmly set. The following chapters will deal with several of these events, like, for example, the events at Roswell, New Mexico. They occurred within a fortnight of Kenneth Arnold's sighting but caused little stir at the time following a US Army Air Force press conference explaining the initial 'crashed disc' report as a misidentified weather balloon.

Within five years of the Arnold sighting two new features to UFO cases were well established: overflights of huge numbers of objects; and radar-visual sightings. One brief flap contained both. On the night of 19/20 July 1952 Washington was buzzed by at least eight unidentified objects which flew into restricted airspace over the White House and made themselves scarce as jet fighters belatedly arrived. Exactly a week later a repeat performance was staged with more targets and, briefly, an air force fighter was apparently surrounded in mid-air as the frightened pilot radioed for instructions. The publicity was huge and the official explanation, that the pilots and radar personnel had simply misidentified the effects of temperature inversions, impressed no one.

This case boasts witnesses on the ground, radar operators convinced they were dealing with real targets, military pilots unable to push their machines to match the airborne objects and civilian pilots diverted to see the same objects. Radar tracks suggested speeds of the unidentified targets of up to 7,000 mph, a performance beyond the best military aircraft today. There is also clear evidence that Captain Edward J Ruppelt, then in charge of an official US investigation into UFOs, was actively misled and hindered during the events.[24] On the face of it, this case is a classic. It remains unexplained and some of the evidence, like the reported speeds of the objects, suggests no conventional explanation is possible.

Predictably, there is some scepticism. More importantly, the Washington overflights introduce other elements of UFO lore that are now an accepted part of the whole business. Namely, the conspiracy and cover-up angles. It is a standard belief amongst many that the US government has knowledge of UFOs, and has possibly agreed with those operating the craft to trade their technology in exchange for non-interference in the abduction of US citizens. The Washington case has echoes of conspiracy. Edward J Ruppelt was warned that a major UFO event was likely to happen before the Washington case. At the time he was head of Project Blue Book, an official US government investigation into UFOs, and yet he was hindered in investigating the case. The implication is obvious and is supported by many events and much evidence since 1952. There is a level of secrecy above those officially acknowledged to exist and some UFO investigation is handled at this level.

Clear evidence of cover-up does not prove the existence of aliens. Some research links the sensational UFO stories of the 1950s with the US government's attempt to use the hysteria to recruit the general public as both official and unofficial sky watchers. The cold war was escalating and there were serious gaps in America's fledgling radar coverage. If this is true, it was an inspired move.

The Washington overflight is, arguably, the first of another type of UFO case: the 'holy grail' report, which presents what appears initially to be the perfect case. Something substantial enough to convince sceptics and significant enough to suggest the involvement of a genuine alien intelligence. Recent examples include the alleged crash of a gunned-down UFO in Botswana; the 'Manhattan Transfer' case in which a woman was reportedly floated out of a New York skyscraper in full view of the Secretary-General of the United Nations; and the film of a supposed autopsy on a dead alien. Granted, the events in Roswell predate the Washington overflights but anyone familiar with the UFO literature of the 1950s, 1960s and mid-1970s knows that Roswell is conspicuous by its absence most of the time so – by default – the Washington overflights are probably the first 'holy grail' case. The general trend amongst such cases is a spectacular launch on the world, a period of controversy and the eventual acceptance by most interested parties that the whole thing is a hoax. Of all the seemingly perfect cases the Washington overflights are the hardest to break. If they are genuine as written then they support the notion that the US government cut a deal after the UFOnauts demonstrated their superior power.

Other evidence includes: the odd feature of the two Saturday nights; the coincidence that on the first night the nearest military air base was inoperative because of runway repairs; the perfection of the case as a media-feeding frenzy; and the apparent prior knowledge of a select few which suggests a covert but earthbound operation. If so, this would be in a great military tradition. For example, the Stealth Bomber was tested against civilian radar. When it vanished within one sweep of a radar beam, generating UFO reports, the military knew the technology could work against an enemy. But, if the 1952 Washington flap was a covert operation, it was one hell of a stunt. Despite so many people being involved, there has been no major leak of information to this day. Also, some of America's best radar facilities, both civilian and military, were fooled by returns moving at incredible speeds. Then again, if it involved craft from outer space why haven't they staged a similar display since?

Less credible, but much more fun, are the 1950s contactees, a bizarre and colourful bunch who claimed direct contact with aliens. Most of the contactees appeared to meet B-movie characters who combined a surprisingly human appearance with a generally benevolent line in anti-nuclear rhetoric. George Adamski remains the most celebrated of this crowd. His monster-selling work *Flying Saucers Have Landed* (1953) details his meeting with a man from Venus. Adamski's celebrity led him, apparently, to a papal meeting. It also led to further books and a die-hard following, loyal to this day.

If Adamski achieved the greatest market share, the most significant developments linked to contactees are probably those that took George King from a London taxi driver to a multi-titled religious leader preaching a complex cosmology. King claimed a 1958 contact on Holdstone Down in Devon and gathered a band of followers who continue to thrive after his death. Many in the UFO fraternity mock the beliefs of The Aetherius Society and, at face value, they do seem incredible. They say that all the planets in our solar system are inhabited, mainly by beings who exist in realities beyond the detection of our science. Jesus is alive and well and was the being who met King in 1958. King was subsequently involved in interplanetary conflict and helped to save our planet. Chanting Buddhist mantras en masse can charge prayer-powered batteries which can subsequently beam their energies to prevent conflict and disaster. Ridiculous or not, King's Aetherian followers maintain their headquarters in Fulham, present a media-friendly face to the world and are obviously happier and more fulfilled for their involvement. The foundations of the group may see them compared to other, more dangerous cults but, in my journalistic experience at least, The Aetherius Society is probably the most helpful and sincere UFO-related organisation I have ever approached. I even spent a very agreeable afternoon with some of them chanting Buddhist mantras on Holdstone Down. And, before you ask, I don't share their beliefs.

Truman Bethurum's book *Aboard a Flying Saucer* (1954) is a connoisseur's contactee tale. His romance with Aura Rhanes, who is 'tops for shapeliness and beauty', is classic pulp science fiction, presented as fact. As is *Flying Saucer From Mars*, the account of a 1954 meeting between the mysterious Cedric Allingham and a Martian on a Scottish beach. Complete with blurry photographs of the Martian and his flying saucer, the book reads like a blatant parody of Adamski's bestselling story. Allingham's book retains an interest amongst the UFO community where it is widely accepted that television astronomer Patrick Moore was involved in its creation, motivated by a wish to prove the credulity of the flying saucer fan base.[25]

From the end of the 1950s more serious contacts were being reported. It is debatable when the era of the abductee began. In recent years, many abduction reports have been collected and many historically strange disappearances, from the men of the Flannan Isles Lighthouse to Moses's gathering of the Ten Commandments, has been claimed as an abduction by someone.

In 1958, Antonio Villas Boas reported that he had been abducted from a farm in Brazil. His story involved coerced sex with an alien female who indicated through gestures that she would have his baby. In 1961, an American couple, Betty and Barney Hill, were chased in their car by a bright light, experienced missing time and subsequent nightmares, and eventually underwent hypnosis. At this stage of the investigation memories of an abduction experience were unearthed. The therapist involved, Dr Benjamin Simon, was asked if he thought the memories were real and replied, 'absolutely not'.[26] However, there is a claim the object the Hills saw was tracked on radar[27] though this was only one radar set to guide incoming traffic to the nearest airport. The radar system covering the wider sky picked up nothing of significance to the case.[28] The Hills' case became public knowledge before the Brazilian event and is, therefore, accepted widely as the first modern abduction. Either way, by the mid-1960s, there were complex and sometimes contradictory reports of human beings being taken out of their environment by aliens who appeared to have a distinct interest in our reproductive capabilities. Then the debate began about whether this was happening in the minds of the witnesses or in reality. The complex argument, which is fuelled by many more investigations, still rages and we'll see more of that later in the book.

So, the concept of UFOs covers everything from the more vivid parts of the Bible to the possibility that aliens are already amongst us. The experts don't agree about any of it. We lack definitive answers because ufology is an investigation that still needs to find itself. One of these days we might be smart enough to ask the right questions.

The Evidence for Alien Invaders

UFO investigators do not study UFOs. They study reports. The best reports make the kind of cases we have already begun to outline. They lead people to believe in particular explanations. In most cases this belief is a matter of faith. The faith runs deep. Differing opinions have been known to result in the hurling of both insults and writs on a regular basis. Arguably, this has deprived ufology of the kind of wide-ranging investigations that would genuinely shake out answers. One effect of this infighting is that sceptics are convinced that there is nothing to investigate.

Alongside writing for a living I spent decades teaching undergraduates to write for a living. Amongst colleagues teaching a range of Higher Education courses you would expect some enlightened thinking. But, when I was writing the first edition of this book, I was talking to one colleague, discussing my latest writing projects. 'Oh neat,' she said, 'They pay you to write a book on something that doesn't exist.'

'Have you ever seen anything you couldn't explain in the sky?' I asked.

'Well... yes, but...'

'But what?'

'There's no evidence, is there?'

Wrong! There is plenty of evidence. The problems start when we ask the next question. Evidence for what?

The 'no evidence' belief generally comes attached to another thought: that UFOs mean aliens. There is no hard, incontestable proof of alien existence. There is much faith that such proof exists and many people think this proof is in the hands of governments, especially the US government. The evidence for this belief remains the most lucrative material in the UFO world and amongst the UK's bestselling UFO books ever are Timothy Good's *Above Top Secret* (1987) and *Beyond Top Secret* (1996)[1]. Their appeal rests on building case studies into an argument claiming that a range of hard-to-explain events leads us to one conclusion, i.e. the fact we have proof of the existence of aliens has been covered up.

The problem with this argument is that proving alien existence is extremely hard. This is true in both the popular and scientific domain. In scientific terms it borders on trying to prove a negative, i.e. that something did not originate on Earth. The book, or reading device, in your hands now could have originated on another planet.[2] Scientifically you could only prove the make-up of the constituent parts and use other information like the publisher's address to trace its origin. All of which might constitute acceptable proof in a court of law. But it could still be fabricated. In which case the evidence you needed for alien existence could be in your hands and disguised as something else.

There are those who would claim there is hard evidence of alien existence that is misunderstood in a similar fashion. If you have bestselling books like those mentioned, access to the internet and key terms like 'Area 51', 'Implant' and 'Men in Black', you may investigate these ideas in real depth. Six months later, armed with the same terms, you could still be finding new and lively little corners of the internet in which some variant of a conspiracy argument was being stoked by a dedicated coterie of message posters. You would, however, not have come across hard definitive proof that would convince a non-believer.

So, at risk of offending those who believe in the proven existence of aliens, and also at risk of disappointing those of you who want spectacular secrets in return for the cost of this book, I'd like to suggest another route through the evidence. Look at the different threads, consider what we know for sure and wonder where it might lead us. If that is a disappointment, please remember what I said in the introduction – that this is a subject in which an amateur investigator can still strive to make the most incredible discoveries and the most active mind can find a lifetime of lateral thinking. If you want discoveries that will shake the world, you could do worse than study what you'll find in this chapter and get involved yourself.

Hard evidence comes in many different forms. Physical objects, mysterious traces, useable data for social scientific investigation and, occasionally, the truly inexplicable.

The Hard Stuff

The really hard evidence consists of things you can see, feel and hold. There is no shortage of claims relating to such evidence. One area of belief suggests that we have all, indirectly, handled such hard evidence. For my money, the original *Men in Black* (1997) is one of the best UFO movies ever made, not least because of its solid roots in UFO lore. At one point a stunned Will Smith asks how a space port is funded. Tommy Lee Jones casually explains that a rake-off from back-engineered alien technology ('microwave ovens, Velcro, computers') is sufficient. One strand of UFO belief claims that back-engineering is there for all to see. Back-engineering is a term you won't find in a standard dictionary. Like many UFO-related phrases it has a loose meaning, crudely equating to stripping down advanced technology, understanding it and recreating it with our own resources.

Forget the microwave ovens for a second. UFO-related tourism is an under-appreciated feature of the present world. Just ask my wife where we spent our honeymoon! One hot spot is Area 51, Groom Lake, Nevada. A lonely expanse of dusty desert and dry mountains to which a dedicated band has laid siege for several years. The result has been fleeting glimpses of airborne lights performing manoeuvres which appear beyond the capabilities of all known aircraft. In recent years the security zone around the facility has been extended. That hasn't deterred UFO tourists or filmmakers – like the team making *UFO Hunters*. Videos and photographs have been widely reproduced and all shades of UFO opinion would probably agree on some of the main points about Groom Lake:

a) The base is the site of testing for future generations of US military hardware.
b) Security personnel and restricted access to certain areas are there to protect the secrets under development.
c) The objects observed and videoed are hard, physical evidence supporting the above statements.

To the people seeing and interpreting these sights the objects are UFOs. Literally, unidentified and flying. UFO literature, websites and conferences drown in claims that this site is where alien technology is back-engineered. A number of people, notably Americans Bob Oechsler and Bob Lazar, have come forward to claim direct involvement in these projects. Lazar's claims include his involvement in attempting to replicate an alien propulsion system with terrestrial materials. Lazar claimed the system was based on an antimatter reactor. In the face of official denials of his involvement Lazar produced some hard evidence of his own in the form of a payslip and internal phone directory to indicate he was inside the facility.[3] His reputation hasn't been helped by a conviction for pandering (a variant of pimping a prostitute) or another conviction for transporting restricted chemicals over state lines. Lazar's claims are further undermined by the fact that his company was fined for offences involving the shipping of chemicals and equipment used in the manufacture of illegal fireworks. Lazar matters here because in the face of this he displays another quality noted in the UFO community – the ability to remain resilient and marketable. The 2018 documentary *Bob Lazar: Area 51 & Flying Saucers* restated many of his original claims and found an audience willing to believe them.

All of which proves – at least – that Area 51 remains sensitive from a security point of view, a compelling subject for many and a revenue earner for those prepared to produce media content on the subject. Opened in 1954, the base has been home to secret projects like the Lockheed SR-71 Blackbird, the U-2 spy-plane and the B-2 Stealth Bomber. To date, the hard evidence proves only that the technology on site starts somewhere around the cutting edge and aims for a realm way beyond that. Exciting for sure but the evidence of alien involvement still rests on the reports and reputations of those who claim to know.

The alleged alien implants don't look anywhere near as exciting as a B-2 Stealth Bomber but they have fallen into the hands of ordinary ufologists and rank-and-file laboratories. Implant lore is complicated. The crude claim is that some abductions involve the forcible insertion of small objects into those abducted. Evidence for implants includes X-ray images and supporting medical testimony. For example, a common location for implants in humans is in the upper reaches of the nose. Some of those allegedly implanted have reported spontaneous nosebleeds. The claimed purpose of the implants is monitoring physical and emotional reactions and tracking the abductee. The ways in which these small and varied objects achieve this aim are harder to establish.

A few alleged implants have been surgically removed. The retrieved objects have been examined with predictably ambiguous results. The parting shot in a statement from one investigator provides the perfect summation to the problem. In his bestselling *Abduction: Human Encounters with Aliens*, Harvard psychiatrist John E Mack discusses the analysis undertaken on an implant retrieved from the nose of a young woman. An analysis proved the presence of – amongst other elements – carbon, silicon and oxygen. A nuclear biologist told Mack that the item was not a biological specimen but 'could be a manufactured fibre'. As Mack succinctly states, 'It seemed difficult to know how to proceed further.'[4]

Difficult for several reasons. Firstly, there was no proven alien artefact with which to compare the implant. Secondly, the implant was an anomalous collection of elements, making any kind of positive identification a matter of luck and guesswork at best. Most implants end their short stint under the microscope in this way. They are identified only in terms of their make-up. Their real purpose and origin are a matter of conjecture.

In favour of the notion that this hard evidence amounts to alien intervention are several circumstantial points. There is some consistency in the location of implants, with the upper reaches of the nose being a particular favourite. Time and again the implants are impossible to trace to any specific earthly origin. They could be a manufactured fibre. But whose manufactured fibre? A number of medical professionals have been willing to stand up and support the belief in implants. Dr Roger Leir allowed cameras into the operating theatre as he removed alleged implants. This process proves that some of the objects in question are genuinely retrieved from inside human bodies. His media work included the book *The Aliens and the Scalpel* (1999) and the documentary *Patient Seventeen* (2017). Leir died in 2014 but, for the final 20 years of his life, he was a prominent advocate of the alien implant theory, speaking at conferences and writing prodigiously. He earned himself notoriety but never convinced sceptics like Joe Nickell who suggested Leir's collection of random objects could be explained as little more than assorted bits of domestic shrapnel and splinters.

Others share Nickell's view that the hard evidence can be read differently. Since the objects appear to be made of elements found readily on Earth it is entirely possible that many, if not all, are accidentally taken into the skin and remain there harmlessly. The 'Strange Days' section of *Fortean Times* magazine runs many medical curiosity stories, some of which involve people who have remained blissfully unaware of objects as large as nails in their bodies for years. There is also a clear possibility of fraud in some cases. Implant fraud in search of notoriety or simply a thrill isn't such a far-fetched idea. The things that people will do to themselves and their bodies push the bounds of believability. One doctor in Israel amassed an astonishing collection of objects removed from the rear ends of his fellow human beings whilst *Fortean Times* editor Bob Rickard once told me a story about a man who repeatedly swallowed a boiled doll's head. This was the same head, swallowed, passed and boiled time after time. There is a difference in motive between the auto-erotic creativity of these fun-lovers and the alleged implant victims. There's also a tonnage of eye-watering clickbait out there detailing an endless array of random animal, vegetable and mineral matter found in human rear ends. So, the point stands, some people do things to their bodies that the rest of us would never contemplate. In such company, those wilfully deceiving themselves and others about the origin of a strange sliver of metal under their skin may be some of the milder cases.

In addition to the implants there have been laboratory tests carried out on alleged debris from crashed craft. The belief in crash retrievals is a major part of many general UFO tomes. The stories are generally set against a wider scenario in which governments gather information about the true purpose of alien races through the accidental acquisition of craft and their occupants. Most of the evidence for such stories begins and ends, like the case of Bob Lazar, with tantalising claims and secondary evidence in the form of drawings. A selection of items of dubious authenticity have been advanced over the years as claimed fragments of extraterrestrial craft. As with splinters of the true cross and other religious relics, the main reason to doubt such items is their somewhat contradictory nature. There is no shortage of alleged alien artefacts made entirely from materials easily found on our planet and despite detailed claims about the incredible properties of some material linked to UFO events – notably the properties of the fragments of the craft reported to have crashed at Roswell – nothing has been examined in a laboratory to date that supports the more incredible claims.

There are some fragmentary clues to alien incursions into our airspace that are impossible to refute. Perhaps the best comes from a case dubbed 'The Siberian Spacefall' by Jenny Randles.[5] A burning aerial object stunned the scattered population of Siberia on 30 June 1908 before detonating over a remote forest area. Of all the reported UFO retrieval cases this one event has generated by far the greatest amount of hard physical evidence. On the day of the incident the fireball and heatwave that followed the explosion were experienced miles from the impact. A shock wave circled the Earth twice and damage to property extended almost 400 miles beyond the impact. By 1945, some of the scientific establishment noticed the clear similarity between photographs of the aftermaths of Tunguska and the American nuclear weaponry unleashed on Japan. At the start of the following year, Alexander Kazantsev, a Russian author and ufologist, was first into print with an idea that is still attached to this case. In a short story, he suggested that an alien spacecraft powered by nuclear engines had exploded over Siberia.

Arguments supporting the exploding spaceship theory draw some strength from the inability of science to agree on what is proven by the mass of evidence from this event. Scientists have put forward a number of theories including the intrusion into our atmosphere of a dead comet, a meteor or an asteroid, all of which are supported by substantial parts of the evidence. The fact that the collection of aerial photographs of the event were burned in 1975 on the direct orders of Yevgeny Krinov, Chairman of the Committee on Meteorites of the USSR Academy of Sciences, has fuelled conspiracy theories. Though a prosaic explanation, that this was a safety initiative to dispose of hazardous nitrate film, also makes sense. The Tunguska area was drenched in evidence, but the most significant investigations of soil, water and rocks only took place years later. At the time a shockwave from the blast was detected across Europe and the skies as far away as

Scotland glowed at night. Amongst the evidence gathered were tiny debris including silicate and magnetite spheres found in soil samples. These debris resemble fragments, called tektites, which result from the fusing of sand and rock during the intense heat of a nuclear blast. The most reliable theories suggest an airburst for an object at least three miles above the ground. Some eyewitness reports stated that the sound of the object was heard at the same time as it passed overhead. If so, it was moving below the speed of sound. Slower than a meteor or comet. 'The Siberian Spacefall' remains a staple of UFO history.

Tantalising Traces

Aside from the hard evidence, the most commonly advanced, definitive proof for the existence of UFOs or aliens is that which leaves some kind of trace. Photographs, films and videos make the most spectacular evidence of this kind. The most convincing cases in this section are those which involve radar and evidence on the ground.

Photographic evidence could prove the case for extraterrestrial existence beyond doubt if the photographs themselves could be proven genuine. From the clear pictures of George Adamski to the dismemberment of the alien under autopsy in the infamous film first screened in the 1990s, this evidence has started its public career spectacularly and quickly found itself reduced to a hard core of supporters. The features that link much of this evidence are the murky origins of the material and the inability of the owners to clear up the mess.

A handful of UFO photographs and films continue to prove perplexing. Foremost amongst this collection is, arguably, a series of photographs taken by Brazilian photographer Almiro Baraúna off Trinidade Island (not Trinidad as sometimes reported). On 16 January 1958, a Brazilian Navy ship was moored off the island when a saucer-shaped object circled the sharp peak and sped out to sea. Baraúna's famous pictures show the object tilted at an angle. Blow-ups of the shots show a grainy texture to the object, which was witnessed by around 50 sailors. The pictures of the object resemble the planet Saturn, squashed and elongated. Reports from the ship also suggest a series of electrical malfunctions in the aftermath of the incident. In this case we have a series of consistent photographs, showing the progress of a mysterious object, with the negatives, the date and time of the pictures and the location itself in absolutely no doubt. The captain of the ship hosting Baraúna insisted the photographer develop his reel immediately and strip to his swimming trunks to avoid any possibility of his swapping an already exposed roll of film for the one in his camera! The pictures were a staple of UFO articles and debate for years. Online coverage of the incident today sometimes includes a statement from Baraúna detailing the events and number of witnesses but there has always been some scepticism suggesting that either a misidentification or a simple double exposure explains the image. After the photographer's death in 2000, one of his friends, Emilia Bittencourt, came forward to claim the photo was double exposed, stating: 'He got two spoons, joined them and improvised a spaceship, using as background his home fridge.' After which, she claimed, Baraúna loaded the previously exposed film and produced the final image by re-exposing the crucial frames as he shot pictures on the deck of the ship. Bittencourt's story matches some criticism of the original photos because the changes in the sky behind the object seem to be considerable in light of the claim that all four crucial images were shot within 20 seconds.[6]

Oregon farmer Paul Trent kept his pants on and took two photographs of a large object that passed his property on 11 May 1950. The generally scathing Condon Committee, a US government-sponsored investigation into UFOs, cited the pictures as the best photographic case in its files and, in 1975, Dr Bruce Maccabee, an optical physicist for the U.S. Navy and prolific UFO researcher, concluded the images showed a real physical object with the pattern of shadows and light in the image suggesting it was a large object at some distance from the camera. Predictably, there is an equally robust body of investigative evidence contradicting the claim and more recently a number of the usual sceptical suspects have poured effort and computer technology into the investigation to produce an evidence-based argument which suggests the photos may show the wing-mirror of a truck suspended on a thread. Robert Shaeffer – whose book *Bad UFOs* sets out to demolish several sacred saucerian cows – has been foremost amongst them.[7] None of this has deterred many within the UFO community from their continued support for the case and Paul Trent and his wife Evelyn (who both died in the late 1990s) insisted to the end of their lives that their pictures were genuine.

Films and videos of alleged UFOs also have their hardcore of classic cases. One of the most celebrated is the controversial multiple-video sighting surrounding an eclipse seen from Mexico City in July 1991. Many residents of the city recorded the eclipse on their camcorders and a virtual UFO mania took hold when well over a dozen recordings showed a stationary aerial object. At first glance this case was perfect with multiple recordings made from different angles at the same time. The eclipse, clearly evident in the recordings, establishes date and time beyond question. The case soon set the UFO magazines and embryonic internet coverage alight. There is no doubt that the recordings show something hugely interesting but a sceptical investigation has calmed the furore surrounding this seemingly incredible case by suggesting that the videos show two objects. One is almost certainly the planet Venus, visible during the day because of the eclipse. The other is large, distant and apparently stationary. It could be anything, up to and including an alien spacecraft. It does, however, look and behave very much like a weather balloon.[8]

Few filmed objects come from radar-visual cases. One of the most durable cases in the literature took place on 20/21 December 1978 when a television crew from New Zealand filmed airborne lights from on board an Argosy freighter aircraft. The 23,000 frames of 16mm colour film devoted to the Kaikoura lights have never been satisfactorily explained. Much of the media furore at the time was overplayed. For example, one frame in which a light appears to perform an incredible looping manoeuvre is almost certainly down to camera shake. For the most part, the film shows the work of a rambling and excited reporter and a cameraman who is struggling with a vibrating plane, a cramped cockpit and equipment demanding his constant attention. What they filmed were distant and small-scale lights in the sky. Poor focus made the lights appear larger than they were to the eye and Channel 10 reporter Quentin Fogarty adds a few cod dramatics to the proceedings with his commentary. The line 'Let's hope they're friendly' was later the title of his book on the subject. The recording matters because similar lights had, apparently, generated eyewitness sightings and radar reports in the preceding days. Radar reports appeared to match the sightings as they were filmed and later flights with better recording equipment on the same route generated nothing to match the original film.[9] New Zealand's Ministry of Defence eventually concluded the sightings were caused by the bright lights of squid boats reflecting off low cloud.

Recently a selection of gun cockpit footage from US Navy aircraft has generated great excitement, apparently showing small objects exhibiting sufficient speed and manoeuvrability to excite the pilots encountering them (a point emphasised by the running commentary from the pilots in some filmed moments). Two pilots – Chad Underwood and David Fravor – went public with an incident that occurred over the Pacific Ocean in 2004. Fravor – the incoming pilot who had completed his mission – had seen a strange object. He would later tell the *New York Times*, 'The thing that stood out to me the most was how erratic it was behaving. And what I mean by erratic is that its changes in altitude, air speed, and aspect were just unlike things that I've ever encountered before flying against other air targets.' He briefed Underwood who was about to take off. After encountering a blip on his radar Underwood managed to film something strange. The film, made available in 2017, was confirmed as genuine by the Pentagon in 2019. Excited aircrew and fast-moving points make an impressive combination when watched over a short period. Some, however, aren't convinced and a *Skeptoid* podcast on 1 May 2018 on 'The Pentagon's UFO Hunt' took some of the footage to task, suggesting the UFOs in vision were a known camera effect: 'In short, what's happening is that the stingray shape – two roundish wings with a short tail – is how any single sharp point of heat appears through the glare filter of the FLIR pod mounted to the fighter plane. Other confirming examples of this shape are widely available online.'[10, 11] Not all the cockpit footage shows a stingray-shaped object and the debate rages over what exactly appears on the films. But a final point applies to all twenty-first-century films and photos. Special effects that were once the preserve of a highly skilled minority are widely available today, including on smartphones. Predictably, the temptation to create work capable of a short and spectacular viral success appears to afflict much of the world's population. Consequently, amusing clips, like the 'alien' caught on CCTV in La Junta, Colorado, gather views and comments rapidly.[12] When ufologists first began investigating images of alleged UFOs, the supply of researchers was sufficient to deal with the incoming evidence. Now social media has tipped the tonnage of evidence into overdrive. Though, taking the La Junta case as an example, it is teaching us something else. The different shades of opinion reveal themselves in the comments posted on sites like YouTube and the level of general interest is clear from the number of clicks given to any clip.

Some other film, never seen by the public, is known to exist. Gun camera film from RAF aircraft was seen by MOD man Ralph Noyes. He believed that the armed forces knew that UFOs were real. His opinions and the cases that formed his view are coming up later in the book. Also coming up later are the investigations of Project Hessdalen in Norway which filmed airborne lights moving in a remote valley and recorded radar traces of the same objects.

In addition to traces on film and radar, there are several cases in which people and the planet have also been left with a lasting imprint. Landings and direct contact with humans have left burn marks, radiation traces and other damage. Much of this evidence is inconclusive but a handful of cases are spectacular and significant. One of the best concerns the experience of Stefan (Steve) Michalak at Falcon Lake, Manitoba, Canada on 20 May 1967. Michalak, an amateur geologist, approached a landed craft. A door opened and closed on the object as Michalak spoke to the occupants, whom he could hear conversing in some strange language. A hatch then opened on the craft and Michalak was blasted by hot gases, sustaining significant burns. A long and painful scramble to safety ended with him in hospital where a host of medical complications ensued. Chest burns formed a grille pattern on Michalak and a range of other symptoms, including imbalances in his blood, suggested some exposure to radiation. More than 20 doctors struggled to find any plausible explanation for health problems that continued for months. No diagnosis or conclusive proof of exposure to radiation was possible and the long-term effects included skin problems and a propensity to blackouts. Other people have also reported health problems as a result of UFO sightings. Better known by the name of its location, Falcon Lake, Manitoba, the case continues to impress and bemuse people, though a body of sceptical opinion sees it as a blatant hoax, suggesting most of the story was fabricated by the key witness and the injuries were self-inflicted. The Iron Skeptic pulls no punches: 'Give me a potato masher and a campfire, and I can duplicate what happened to him.'[13]

UFOs have left traces on the ground. In many cases the late arrival of investigators has limited the credibility of any investigation into such evidence. However, in France there has been cooperation between scientists and the government to establish an investigation bureau. This was set up as GEPAN in 1977 and reconstituted in 1988 as SEPRA before becoming GEIPAN in 2004. The role of the various incarnations of the organisation has varied. SEPRA took on the responsibility of monitoring events like the re-entry of spacecraft into French airspace. The current organisation's wide-ranging remit involves the investigation of all unidentified atmospheric phenomena and, since 2007, their records have gradually been made widely available. This organisation has established a practice of working with others, including the gendarmerie, as necessary. One case in particular is worth considering here. On 8 January 1981 at Trans-en-Provence a landing was reported by 55-year-old farmer Renato Nicolaï, who was busy building a small shed. By the standards of many reports this was a mundane event but the traces on the ground were swiftly examined by a GEPAN team who used laboratory soil analysis, with control checks, to establish small but definite changes in the soil. Amongst the evidence, verified by analytical chemists at two universities, was the unexpected discovery of a high number of negatively charged ions and evidence of carbon polymers in the soil. Predictably the more sceptical elements in UFO research have criticised the thoroughness of the investigation. The results from the investigation prove only what was sampled in the aftermath of a report. Nicolaï's initial thought was that the object was a piece of military hardware from a nearby base. Interestingly, the main witness confirmed that the event would have been visible to motorists on a nearby road, but none of them came forward to add a witness statement.

Much of GEIPAN's work has also demonstrated a high degree of critical thinking and turned up evidence that shames other, less scrupulous investigators. One vivid ring in a field, believed to be a landing trace of a UFO, was shown to be fungal. Perhaps the most entertaining case in the files involves a crater which appeared overnight in a farmer's field. Meticulous investigation revealed an underground detonation of a 50-year-old Second World War bomb, even tracing its manufacture to England.[14] Burns, markings and residues from alleged landings are plentiful in UFO history. The vast majority of UFO cases are investigated by amateur groups with minimal funding and nowhere near the same ability as GEIPAN to access professional help.

Elsewhere, the claims border the very edge of believability. The Massachusetts Institute of Technology hosted a high-profile symposium designed to gain academic kudos for UFO and alien abduction investigation. Journalist CDB Bryan listened in astonishment to claims that aliens had successfully treated AIDS and colour blindness.[15] For years conferences regularly heard reports of unexpected pregnancies, confirmed by doctors, which had spontaneously ended. The belief, supported by the claims of leading abduction researchers like Budd Hopkins, John Mack and David Jacobs, is that alien abductors are removing pre-term foetuses to allow them to continue their existence on their craft. Their aim, apparently, is to interbreed with humans and produce a race that may eventually be involved in changing the direction of life on Earth.

Elsewhere, hard physical evidence takes the kind of bizarre twists that will be all too familiar to you by this stage of the proceedings. Like many with an active interest in UFOs I have had my hands in Bob Taylor's trousers! Superb trousers they are. The rugged work trousers of a Scottish forestry foreman. Bob and his trousers became estranged around the time his experience in a forest near Livingstone, Scotland, became a celebrated UFO event. The trousers have become a holy relic of UK UFO investigation and have spent most of their life in the possession of Malcolm Robinson, a UFO investigator who took part in the initial investigation of Taylor's claims and wrote the only comprehensive book devoted to the case: *The Dechmont Woods UFO Incident* (2019).

On 9 November 1979 Taylor encountered a large Saturn-shaped object in a forest clearing. He described this object as fading in and out of view. At times he could see right through it. Two smaller globes with protrusions, similar to Second World

War sea mines, flew towards him at which point a rapid series of events left the forester unconscious and his trousers ripped. Lothian police would later treat the incident as an assault, generating a case file and making the incident a rarity amongst British UFO cases. Like Stefan Michalak, Taylor sought medical attention and struggled to make sense of an experience he didn't want. He underwent a hypnotic regression in which he recounted the whole bizarre experience exactly as he had reported it at the time. Some objective investigation was brought to bear on the evidence. The famous trousers showed outward tears as if ripped mechanically and indentations on the ground corresponded with Taylor's report of the objects seen. No additional tracks into and out of the forest were found to indicate that any conventional forest machinery could have made the indentations at the site of the incident. Tests for other evidence like radiation and residues, however, proved negative.[16] Malcolm Robinson's book-length investigation of the case devotes over 90 pages to a consideration of 16 theories from the obvious (genuine ET event/all a hoax) to the plausible but highly unusual (Bob Taylor was unwittingly under the influence of Atropa Belladonna).[17]

Taylor's experience is one of many UFO case histories in which odd physical evidence has been combined with an eyewitness report. Some of the best known are the 'car stop' incidents in which petrol engines (not diesels) are reportedly stalled by nearby UFOs. Elsewhere, power failures and power surges have been reported. In one notable British case from 1978 in Risley, Cheshire, Ken Edwards reported a truly bizarre incident involving missing time and damage to a two-way radio in his van which – apparently – resulted from light beams fired his way from a white mass, dubbed an 'entity' in some written accounts of the case.[18] Investigation showed the radio had burned out after an unexplained power surge. Elsewhere, damaged homes, cars and gardens bear repeated witness to the fact that people continually experience something out of the ordinary. Any sceptic suggesting such damage was inflicted by the witnesses in search of publicity is missing the point by several light years. On one occasion I dealt with a man reporting something similar to the Ken Edwards incident. The witness in question was convinced he'd had a genuine and frightening experience. He didn't want to go to the papers. In fact, one of his first requests in dealing with UFO investigators was that nobody mentioned the event to his wife! This witness, like Bob Taylor and the pancake-gathering Joe Simonton, had encountered something that had shaken his belief in all he knew. His sincerity wasn't in question.

Those were a few of the greatest hits in the search for evidence of alien incursion onto our planet. As I told my colleague as she stubbed out her cigarette outside the college, there is evidence, much more than I've had the space to include here. The tortuous 'evidence for what?' question remains unanswered. We have undoubtedly encountered the unexplained. The Tunguska case, for one, almost certainly involves some object entering our atmosphere from space. Elsewhere, radar tracks, unidentified targets and eyewitnesses are occasionally supported by photographic and filmed evidence, and physical traces appear that prove hard to explain. When the best science has been applied to a handful of these events it has strengthened them. Some witnesses may lie and manufacture evidence but other cases produce the most credible witnesses, like Bob Taylor. He cooperated with the investigation, was trusted by the local police and his own GP, and remained resolute and consistent in his account which was supported by physical evidence.

The definitive answers to the questions posed by the evidence here do not currently exist. These cases appear most frequently in the books, videos, magazines and websites that argue for repeated alien contact with mankind. I'd argue this makes good sense in marketing terms but little sense in terms of objective investigation. In a field of investigation packed with claims lacking any evidence and evidence lacking any credence we have a hard core of cases that withstand investigation. They may eventually lead us to discover atmospheric phenomena we don't understand. They may, as we will soon see, continue helping us to a greater understanding of our own psychology. My own feeling is that these explanations and more are behind the cases quoted in this chapter. Perhaps, when we've explained them all away, we'll be left with one or two crowd-pleasers that will have the pro-extraterrestrial crowd screaming, 'We told you so'. If and when that happens, my guess is that the sceptics will counter with an argument that starts, 'Yeah, but you told us all those other dodgy cases were aliens as well!'

The UFO Community

So far, we've explored a brief history and an overview of the best of the evidence. Now it is time to tackle 'the people problem' I referred to in the introduction. Or as I defined it: 'those within the UFO community who gravitate towards the answers they want regardless of the evidence'. In plain English the issue is that much of what we understand as true about UFOs and related issues is really a combination of some facts and some strongly held beliefs. From the point of view of hard, objective scientific fact the 'problem' is that the facts supporting the beliefs don't stand up well enough to prove a point as far as – say – a peer-reviewed scientific publication would demand they do.

The meanings we make in UFO investigation often come from the people involved. The evidence may lead us to opinions but in many cases any kind of conclusion remains impossible. Even the best documented cases, like Tunguska, remain mysterious. Predictably, we have different opinions arrived at by different people. UFO investigators and the community of individuals who remain interested in the subject are most certainly 'different' people. Sometimes that means different as in 'people with contradictory views' and sometimes it means different as in 'not normal like the rest of us'. So, it is time to meet the people. Where else can we start but with the one group to whom we owe the most? The people without whom this book, and this subject, would not exist.

The Witnesses

Very little definitive research has been carried out on UFO witnesses. We have some notions about who these people are and how they experience their events. The first point that should be made about the witnesses is that many of the popular notions that surround them are plain wrong. Most of the witnesses I have met are not in it for publicity, money or any kind of vicarious need for attention. They are normal people who have experienced something abnormal. Most never find themselves on the receiving end of an active investigation of their experience. Fewer still report anything to the media.

No totally trustworthy statistic exists by which to measure the numbers of people experiencing UFO events. In the USA, the Gallup and Roper Poll organisations have systematically included UFO-related questions within their general polling of the population. In 1985, a figure of 6% of the population indicated they had experienced a UFO event. Almost six years later the figure answering the same question had risen to 14%.[1] By 1990 Gallup had polled the American population with this question four times and Roper twice. Gallup's figures consistently came in higher than Roper's for the same question. The rapid rise in belief to the 1990s remained stable and, by 2019, 33% of Americans believed some UFOs to be alien spacecraft with 16% of that population saying it had experienced a sighting. Men were marginally more likely than women to believe in the possibility of alien spacecraft with the youngest section of the population the likeliest to believe and non-graduates being more likely than graduates to hold this belief. The headline discovery of the 2019 survey was – however – that over two-thirds of Americans felt their government knew more about UFOs than it was sharing. The figure was 3% down on the answer to the same question in 1996, but still notably high.[2]

Analysing such raw data and applying a conclusion to the rest of humanity verges on the impossible, especially since America stands out as a country that bothers to survey its population on UFO-related questions. A UFO event in someone's mind may be an obvious sighting of Venus when explained to an astronomer. Jenny Randles once told me something that many investigators have found to be true. Most sightings exceeding five minutes prove to be astronomical events. Crudely, the longer a UFO remains in view the easier it may prove to be explicable because coordinating the details of the sighting with – for example – the position of Venus in the evening sky, might provide a plausible explanation. A good example of this general observation in practice may be the videos during the Mexico City eclipse. The sighting was one in which an object remained in view, was widely thought to be a strange craft and yet yielded to subsequent investigation once the recent release of weather balloons and the position of Venus had been taken into account.

One possible explanation for the rise in UFO sightings and the seemingly high percentage of people who have experienced them may be that a rising population means there are more people around to see strange phenomena when they occur. Another possible explanation, roughly supported by other polling information, is that people have been seeing strange things for years but are increasingly predisposed to interpret them as UFO events. The work of folklorists Walter Evans-Wentz and Katharine Briggs was mentioned earlier. The stories they collected of meetings between humans and fairies of many descriptions overlap with present-day accounts of encounters with aliens in a number of ways. The stories have many superficial similarities but often also contain idiosyncratic twists. Some common factors include time appearing to pass at a different rate to normal for the humans involved, the imparting of wisdom that appears significant at the time but often amounts to very little useful knowledge and the common fact that items taken from the strange realm often prove to lack any strange properties when examined by others. Joe Simonton's pancakes are an example of the last point.

The USA and UK have been home to most of the significant psychological research carried out into UFO witnesses. This research is far from definitive but it does deliver some insights that may form the basis of definitive research in the future. The late Ken Phillips carried out a long-term research project into a range of personal factors of UK witnesses. His reports indicate that many UFO witnesses feel a status inconsistency. Roughly speaking, this means the witnesses experience an awareness that their status in life is out of kilter with their background, experience and/or qualifications. Those who feel satisfaction with their jobs still tend to experience adjustment problems in other areas of their lives.[3] Phillips' investigation, called 'The Anamnesis Protocol', also discovered a high rating for an external locus of control.[4] This indicates that a higher than average percentage of his subjects felt that significant areas of their lives were totally out of their control. A crude demonstration of this might involve a tricky work situation in which management were attempting to implement unpopular working practices. A person inclined towards an external locus of control may resent this as much as any other member of staff but would be more likely to do nothing other than say, 'We can't do anything'. A large-scale American study of UFO witnesses showed a similar pattern of personality traits to Phillips' UK study.[5]

A more controversial train of thought links UFO experience to the growing psychological literature on Fantasy Prone Personality (FPP), which is a theory suggesting susceptibility to fantasy experiences. The concept of FPPs was first advanced in a research paper in 1981.[6] Since then it has appeared in the more scholarly and less widely read end of the UFO literature. Robert Bartholomew and George Howard's investigation into UFOs and aliens contains a succinct analysis of the link between UFOs and the FPP theory.[7] The idea splits the UFO community along predictable lines. The generally sceptical and serious minds of the academic fringe are genuinely excited by a theory that offers explanations for the perplexing reports of witnesses. Especially so when the theory is applied to the stranger accounts in which elliptical wisdom is imparted by aliens. FPP is – for example – potentially a good fit with a case like that in which Antonio Villas Boas claimed to have had sex with an alien female who then indicated through gestures that she would have his baby. The improbability of humans mating with any other species has always convinced some investigators that this vivid event occurred in the mind of the witness and nowhere else. The witnesses and those most sympathetic to the idea of aliens, known as the Extra-Terrestrial Hypothesis (ETH), are often fiercely resistant to psychological explanations, such as the FPP, which come from social science. Many witnesses are the most resistant of all, claiming, in effect, that a bunch of academics can't tell them what they experienced. A valid criticism for sure. Some researchers with a track record of using hypnotic regression to uncover abduction stories are also conspicuously opposed to the

FPP idea. This much was brought home to me when I appeared alongside Manchester solicitor Harry Harris on a television show. The mention of Fantasy Prone Personality had him pointing out that a doctor had assured him it could not be diagnosed. True, but this misses the point.

The existence of FPP is a theory supported by a body of evidence substantial enough to allow FPP to be the subject of a series of peer-reviewed articles. The alleged existence of Gulf War Syndrome is a useful comparison. The theory in both cases has led to accurate predictions and fruitful research but we're still a long way short of proof and, therefore, diagnosis. In fact, we might never get to a point when either condition is diagnosable in its own right. But that doesn't mean the findings and research are wasted. With FPP, we do appear to be in possession of a useful idea that directly tackles some difficult problems. Superficially at least, FPP fits witness and abduction research very well and leaves doubt about the extraterrestrial origin of some, if not all, abduction reports. There are precedents for the situation in which medical science has gradually explained evidence that was previously understood in a supernatural way. The current understanding of conditions like ADHD and Tourette's Syndrome means that we now use symptoms to diagnose and help sufferers. In the infancy of research into these conditions some symptoms were simply perplexing and a general diagnosis using terms like 'mania', accompanied by the prescription of stupefying opiate medications, was all that could be applied. In the past, some of the symptoms of these conditions were taken as evidence of possession by the Devil.

Many abductees have returned with specific information about the origins and purpose of their abductors. This information has been relayed sincerely to researchers, at which point it has often proven to contradict information gathered from other abductions. Predictions gathered in this way are similarly unreliable. Any comprehensive theory of abductions has to account for the bizarre but sincerely reported encounters of people like Joe Simonton and Alfred Burtoo (whose story is coming later in this book). FPP may begin to explain the general sincerity and the odd way in which the nature of abductions appears to change with popular ideas about UFOs and aliens at any given time. It also provides an explanation that links UFO experiences and the fairy lore of old.

Robert Bartholomew and George Howard spent a chapter of their book, *UFOs and Alien Contact*, comparing the best investigations into FPP with the personality traits outlined in biographies of those experiencing UFO events. The results strongly suggest that UFO experiencers have a greater tendency than the general public towards the traits of the FPP. For example, 16% of the general public and 92% of those potentially exhibiting FPP report experiencing psychic phenomena. The figure for those claiming UFO events is 75%. This figure, however, doesn't include the claimed UFO event as a manifestation of psychic phenomena.

To put this in context, nobody is claiming that the bulk of those experiencing UFO events are mad. It is simply that they exhibit traits that in other areas of life, like creative professions, are a positive asset. The best proof for this, oddly, comes from an internet-published paper attempting to prove that the FPP theory was wrong. The original paper discounted some of the main ideas but the researcher invited an independent evaluator of tests to examine the subjects. The independent evaluator was unaware that the subjects were abductees. In comparison to the general public, the subjects were found to have rich mental lives, weak sexual identities, tendencies to paranoia and caution and impaired personal relationships.[8] This is far from a universal appraisal of all abductees but it does suggest that many of those reporting UFO events fall into a group of the population who exhibit personality traits that shape their lives. To be clear, this is not mental illness, simply traits that help determine the individual and their experiences.

This much has been evident to some commentators for a while. I once had the chance to communicate with aliens. We don't need to bother with the whole story. The gist of it revolves around a UFO research group which had established a psychic channel to a group of almost 1,000 alien super-minds who existed in a form approaching pure spiritual energy within the Earth's atmosphere. Superb stuff, and the most endearing trait of these beings was their willingness to answer questions on anything. I got to ask a few. Some, concerning the existence of God, global warming and other major concerns were given short shrift. One question concerned the promise of Michael Knighton, then chairman of Carlisle United, to get the football team I've followed my entire life into the Premiership within ten years (i.e. by 2002 after his 1992 takeover). I doubted the deliverability of Knighton's statement then. In the end he succeeded in regularly taking us to more relegation fights (including battles to stay in the league) than promotion campaigns. How, I wondered, would these alien super-minds have achieved the Premiership? Their reply still gets laughs in front of a live audience. They suggested baking a cake that contained the wishes of the chairman and feeding it to the players who would then deliver on the pitch. If this truly is wisdom from highly advanced alien super-minds then, frankly, I have a problem with it. A workable recipe for any cake that can turn lower-league cloggers into the equals of £100-million-pound internationals beggars belief. I doubt whether heavyweight boxing or reality game shows could compete in a ratings war with a programme that involved locking half a dozen managers and that cake recipe in one room. I don't, however, doubt the sincerity of the people who passed the message back to me. Within the group was a person who openly admitted to psychic experiences. An admission that fits with those who are prime candidates for FPP. The human channel in question was combining psychic experience and alien contact in the same activity. In the circumstances there are many, me included, who would suggest it is easier to believe in FPP than cakes that turn a carthorse of a centre forward into a soccer god. As an aside, one thing Michael Knighton – ex-chairman of Carlisle United – is remembered for is a UFO experience. Once he admitted this in public the fallout led to him threatening to resign his football chairmanship in 1996.[9]

More recently Robert Bartholomew has written on the phenomenon of small-group panic and considered its possible significance in a few particular UFO-related cases in which small groups of people reported intense and unusual events. Notably the Kelly-Hopkinsville encounter (which involved small creatures besieging a remote dwelling and the family inside frantically fighting them off with firearms) and the Knowles Family encounter (which involved an Australian family encountering a UFO during a lengthy drive across the Nullarbor Plain).

None of the above proves a thing. In the twenty-first century we have a handful of useful research studies into those experiencing UFO events. The common factors identified suggest that many of these people often share personality traits that make them different and, in some cases, border on the dysfunctional. FPP or no FPP, those reporting UFO events tend to set themselves apart from society. In reality, they may have been out on their own before their UFO experiences.

Everyone Else

I've regularly referred to the UFO community in this book. You won't find them in any official listings and they don't live in any one place but I'd defend the use of the term for three reasons.

a) One meaning of community is a group of people united by a common interest.
b) I've been using it for years without once being challenged on the grounds that it is inappropriate.
c) The English language doesn't contain a single word complex enough to describe the real situation in the UFO world and 'community' is simply a descriptive and fairly accurate alternative.

The community in question contains a range of interests. Scientists undertaking serious research. Interest groups like the military who retain their own particular involvement and mingle with everyone else only when necessary. Active researchers and a great many others who buy the merchandise, turn up at one or two major events and follow the whole business from a distance. All sections of this community, especially the casual members, show a pattern of people coming and going on a regular basis. Given the differences in motivation and involvement of such a diverse group it would be impossible to define a single personality type making up the membership. However, there have been attempts to understand and explain this community.

One notable thing about the research into those inclined to believe in the reality of UFOs and aliens is the way the results correspond to the traits identified in witnesses. If we take a brief historical overview we can see the knowledge base slowly building over the last few decades. Steven Resta's master's dissertation in 1975 examined psychological traits of UFO believers. The researcher found a high score for external locus of control.[10] Martin Kottmeyer built on this notion of the UFO audience feeling that their lives were controlled to suggest that the whole community was an example of an evolving system of paranoid belief.[11] This notion was also central to one major British study. David Morris examined the readers of a literature he labelled 'techno-occultism' in his book *The Masks of Lucifer*.[12] His definition took in works on ufology and other paranormal and spiritual literature. Roughly, the kind of material you'd find in the 'Mind, Body and Spirit' section at Waterstones. He found a link between an interest in this material and notions of personal status in the readers. His findings suggested that the apocalyptic claims and reinventions of conventional wisdom in works like *Chariots of the Gods* provided comfort to an audience who were struggling for status in their everyday lives. In the simplest terms, there was a comfort in the UFO realities presented in books because they undermined the hard realities of an everyday world in which members of this audience were often on the margins. This concept is not new. Morris quoted Vance Packard's seminal research into notions of social class and status. Morris linked his view of techno-occultist readers to Packard's concept of a 'limited success class'.[13]

Another classic British study formed a PhD thesis for Shirley McIver at the University of York in 1983. Her examination of those actively involved in the UFO community identified some specific motives and personalities. For example, she looked at the Birdsall brothers, who went on to found and run Quest International, an organisation that led the UK market in UFO conferences and merchandise for years. The brothers' youthful interest in employment as secret agents is one item from McIver's study that their fellow ufologists still quote with some amusement. This amusement may be misplaced. If anything, the Birdsall brothers achieved the essence of their dream with a professional life that included deliveries of secret material, the chance to break incredible stories to the world and a network of contacts around the world who met, often in their native Yorkshire, to discuss what was really going on. *UFO Magazine*, their main publication, ran from 1981 to 2004, eventually closing the year after Graham Birdsall's death. Your view of their life depends on the faith you place in Quest International and their generally pro-extraterrestrial message. But those attracted to the life of a secret agent in their early years are often dreaming of being at the cutting edge in an arena of cover-ups, collusion and controlling the world. The Birdsall brothers and their less commercially successful peers have never done it for Queen and country. Their ilk are, however, involved in a crusade for truth and fighting unsympathetic forces who conspire against them. Some of us might question the claims of many in UFO groups about infiltration, tapped telephones, their insider knowledge about UFO wreckage in secret hangers and a host of other staple stories. We would, however, be idiots to question their belief and sincerity.

By the end of the last century there was enough known about the UFO community and their belief systems for substantial academic papers and wider works to be published. One of them – Jodi Dean's *Aliens in America* – chronicled the development of a movement she saw as one of the few belief systems critical of much government policy to have survived and prospered in the cold war, largely because at heart it held up models of good citizenship and used them to criticise elements of public policy, including the way secrets were kept. In referring to models of good citizenship, Dean is mainly considering the way the social standing of key witnesses is used to add credibility to their stories: Kenneth Arnold was a businessman, Betty Hill was a social worker, policeman Lonnie Zamora was the main witness in another famous case, etc. Dean states at one point: 'Those of us attracted to left-wing causes, to critical positions against political, governmental, and corporate authorities, or maybe just to underdogs in general may feel at home in ufology.'[14] Of course, Dean didn't say these were the only people attracted, simply that this general shaping of the community could be gleaned when you looked at the credible research into what was known of them. The internet has widened the involvement to the point many take part without revealing their true identity or purpose (the latter of which they may not fully understand themselves).

If one thing unites those in the wider UFO community it is this sincerity. Outside of active group members and buyers of any kind of UFO merchandise we have a fringe of those involved because of other interests. Researchers building careers tend to steer away from ufology because of the academic stigma attached to it. There is no widely available research on the psychological make-up of those involved in UFO research who are not members of groups. Given the professional suicide liable to result from such labours we might usefully suggest that these people straddle the line that divides dedication and obsession. It is a crude measure for sure but one lengthy insight into the truth of this suggestion comes from Leo Sprinkle. His autobiographical account of an academic career combined with ufology, *Soul Samples*,[15] details a series of tetchy academic spats and self-justifying soul-searching episodes of his own alongside his groundbreaking work with abductees. Sprinkle was one of the first to pioneer an approach that put UFO experiencers in the centre of investigation and, in effect, allowed them a say in what mattered and how the investigation should proceed. Sprinkle's self-help philosophy allowed him to work with groups of abductees but his motives in doing so bring another dimension to our understanding of those involved in UFO research. Sprinkle is an experiencer himself. In fact, those claiming some kind of experience make up a significant but hard-to-quantify section of the general UFO community. Leo Sprinkle is one example who shows the blurring of boundaries in this area. He is an experiencer, a delegate and performer at conferences, an avid consumer of merchandise and also a researcher involved in the field in his professional capacity.

Experiences of a different kind form the basis for the involvement of others. Canadian neuroscientist Dr Michael Persinger developed experiments to prove a theory he advanced in 1977. Persinger, who died in 2018, believed that electromagnetic energy fields can affect the human brain in a manner that causes UFO experiences. Persinger pursued these investigations in his professional capacity at Laurentian University using custom-built lab equipment. He produced results that suggest his original theory had much merit. Some other researchers, including Britain's Albert Budden and Paul Devereux, engaged on very personal quests based on similar beliefs. This is a dedication summed up by Devereux's apology as he ran over time during a presentation to the Fortean Times Unconvention, asking the audience to stay with him a few minutes longer he stated simply, 'This is the work of a lifetime'. Devereux's work on earthlights is based on a belief that natural energies leaking into the atmosphere can form short-lived airborne lights which ride on magnetic fields. Budden developed an 'electro-staging hypothesis' which suggests that Devereux and Persinger were working along the right lines. Budden also studied electromagnetic pollution from sources like mobile phones and radio masts, seeing this as a potential cause of UFO events.

The likes of Devereux and Budden are ambivalent figures to the UFO community because their theories fly in the face of the popular belief in aliens and abductions. Jenny Randles noted in her *Little Giant Encyclopaedia of UFOs* that Budden had 'left ufology behind to align himself with scientists'. A telling comment because Budden continued to study UFO events but his theories find some derision and limited sales amongst the UFO community. The dedication of figures like Albert Budden is in a great tradition of pioneering science. Such pioneers continue to make a telling contribution to ufology because the scarcity of serious scientific investigation in the field continues to leave open the possibility that amateurs can make earth-shattering discoveries. Budden's work overlaps with other studies into electromagnetic pollution that concentrate on the possibility that our expanding communications network is damaging both our mental and physical health.[16]

Military, atmospheric and aeronautical scientists have followed UFO events for many years. More accurately, they have taken a marginal interest but occasionally added theories and ideas to the field which have influenced ufological thinking. The mutual suspicion between the essentially amateur UFO community and those with careers in the military and scientific worlds remains strong. There have been some attempts to bridge the gap. One of the most accessible is probably Edward Ashpole's book *The*

UFO Phenomena (1995). The book, a greatest hits of UFO events with scientific thought added, provides a potted history of some important opinions. Career scientists and military people may exist on the fringes of the UFO community but their work is influential in making meaning and establishing some of the limits of the claims of the UFO researchers. Ashpole notes that of the hundreds of contactee and abduction cases recorded up to 1995, not one had provided 'anything of acceptable scientific interest – say, something on which one could write a paper for *Nature* or *Science*'. Ashpole also puts some current claims into a hard-scientific context. For example, he quotes Carl Sagan's famous observation when faced with the claims of Antonio Villas Boas to have had sex with an alien. Sagan noted that there was more chance of an elephant mating with a petunia than a human mating with an extraterrestrial. Ashpole's highly readable analysis also provides moments of clarity as UFO claims collide with scientific reality: 'Sadly the scenarios of *Star Trek* are wrong. Spock reports to Captain Kirk that the planet's atmosphere is breathable... in reality, one whiff of that "breathable" atmosphere could end a promising career in Star Fleet Command. And maintaining control of our bowels on Planet-X could be an embarrassing problem never faced by Captain Kirk and his crew.'[17]

Scientific and military views are met with some derision amongst the UFO community. The Search for Extra-Terrestrial Intelligence (SETI) is a project with its own institute in California. A methodical search for evidence of extraterrestrial communications based on radio astronomy, the institute exists uneasily alongside popular views in the UFO community. If the claims of just one abductee are proven then the SETI Institute is nothing short of a multi-million-dollar international embarrassment, and the recent massive upscaling of scientific research into SETI is needless. SETI scientists, by contrast, are generally sympathetic to the wishes of the popular UFO writers and researchers but unimpressed with the quality of their evidence or argument.

There is much animosity between military investigation and the UFO community. Officially many world governments, including those of the USA and the UK, admit to investigating UFOs but state that their interest ends when they have examined the defence implications of any specific case. Popular opinion in the UFO community sees this as spin-doctoring at best and, quite probably, an outright lie. Others, including the late Ralph Noyes, who spent his career in the Ministry of Defence, headed Defence Secretariat 8 (DS8) from 1969 and retired in 1977 with the rank of Under Secretary, claim otherwise. Noyes was active in the UFO community for years and consistent in his stated view that the ministry had UFO evidence which had not been made available to the public. Noyes believed that the MOD had conclusive proof that UFOs were real and that they were not alien spacecraft. He also freely discussed an event in 1970 in which he and others in his area of work had been given a lunchtime screening of slides and gun camera footage dating back to the mid-1950s. He stated the material was, 'on the whole unimpressive'.[18] The real nature of UFOs, he said, remained elusive. But it was, apparently, clear to MOD investigators that some of the most incredible UFO events were being caused by unknown atmospheric phenomena. There is a widespread belief in covert infiltration of some UFO groups by the military or security services and it has definitely taken place. The motivation behind it is another matter. If UFO groups in the UK have unearthed a massive case that has subsequently excited the military then it's hard to identify which one it is. The cases most exciting to the military have been military events first, and come to the attention of ufology later on. Some ufologists have been known to infiltrate military installations; indeed, I once sat through a conference presentation that consisted in large part of slides and a story of just such a thing happening. The ufologist in question was photographed in an underground storage facility next to spare parts for military vehicles. Whilst no major security incident resulted from that escapade the motivation for the military or security services monitoring such behaviour is obvious. The extent to which their marginal involvement in ufology and UFO-related events has prompted them to follow up on cases first reported to civilian groups is much harder to prove. Ralph Noyes was adamant that only three cases had ever really got the military nervous during his tenure. Two of these will be dealt with in detail later.

Considering the UFO community in the light of Noyes' comments, it is possible to see how suspicion and hostility might build between military and amateur researchers. Rank-and-file researchers would be right to believe in infiltration and secrecy. For their part the military would feel justified in what they do. As we will see later, the cases that made the military nervous concerned UFOs behaving in a manner that caused initial fears about an incoming military strike. The possibility that the nation could be panicked into launching a military strike against an unknown atmospheric phenomenon is understandably alarming. It's also damaging from the point of public relations, making it easy to understand why secrecy would be maintained. In this climate the military and security services tend to remain terse when they reply to questions, which – in turn – feeds the ongoing belief presented in books like *You Can't Tell the People: The Definitive Account of the Rendlesham Forest UFO Mystery* (2000) that the secrets covered up by military and security organisations include the reality of alien contact and recovered alien hardware. The titles of the books, films and other content sharing these stories are indicative of the angles they take on the subject. For as long as the official line remains where it is, the current stand-off, which has lasted decades, is likely to continue.

Making Meaning About UFOs

We've spent little time considering the cases in this chapter but what we have here is important. It matters because we need to remind ourselves of some vital points relating to the evidence. Firstly, UFOs and UFO information are very much a commercial enterprise. Most of the information in the public domain is there because it makes commercial sense. The bestselling information, like the puzzling cases and the idea of aliens being present on this planet, gets regular exposure and frequent media makeovers. The dedicated research in academic disciplines like psychology is discussed in a few peer-reviewed journals and given bookshop space only in the largest and most eclectic stores. The unchecked internet is alive with UFO information, most of it in the conspiracy/'they're here' category.

UFO publications, websites and videos strive to create the illusion that they present hard information all the time. In reality, we are often dealing with infotainment, a situation in which the facts are shaped to market needs. The truth can be a casualty. People make their own meanings by deciding on the information they receive. Groups may then gather and support each other's beliefs. Not surprisingly, the most popular beliefs remain popular and the cult areas stay on the fringes. This is a situation closer to popular entertainment than an information-based pursuit. In entertainment some staple products, like soap operas, remain popular and repeat aspects of their winning formulae. In information-based pursuits, like an educational course, there is a more linear structure and the goal is a conclusion or achievement. Those with a UFO interest often talk like an educational course and behave like a soap opera.

Ironically, one person to recognise this situation is Timothy Good, the bestselling author who made a fortune from pro-extraterrestrial and pro-conspiracy writing. Promoting the massive-selling *Beyond Top Secret* in 1996, Good told *The Guardian*, 'If ever there's a subject that needs rescuing from its supporters, it's this one.' Good's comment and the work of pollsters and psychologists would suggest that there may be such a thing as a UFO-related state of mind. More accurately, there may be a popular mythology at work in people's minds. Those inclined to believe their lives are controlled from outside may be the most susceptible. They may also be well represented in the regular audience for UFO material that feeds this very belief. Admittedly, there is a smattering of social science research, most of it based on investigations in the US, but the most comprehensive work is quoted here and does show a consistency of results between studies and between the groups of UFO experiences and the audience for material on UFOs. The most significant thing about this work is the consistency in the key areas of external locus of control. It also shows a general belief in UFOs and extraterrestrial life. Two highly-controversial experiments have added some insight into how these facts might manifest themselves.

At the Anaheim Memorial Hospital, California, in 1977 Alvin Lawson and William McCall conducted a survey using hypnosis. They took 16 subjects, some genuine abductees and some who had been previously screened and selected for their lack of interest in UFOs. Using a series of questions, they made the subjects imagine and recount tales of encounters with UFOs. The experiment found many and distinct similarities between the stories told by the abductees and those with little interest. Both groups were encouraged to see the events as real. In 1993, the British researcher John Spencer attempted something similar and found similar results.[19]

These small-scale studies remain controversial, not least because of the ethical position of leading those with little or no belief in UFOs to believe they have had genuine experiences. However, the experiments do open the possibility that the public at large has a mental template of a UFO experience. This, in theory, could be triggered by an event or some effect on their brain, as suggested by researchers like Persinger, Devereux and Budden. These experiments and the other material on the make-up of those experiencing UFO contacts also suggest that much UFO-related experience may be psychological. This may present a model in which the personalities and their experiences remain the same over a long period and their interpretations vary with the ideas of a particular society. In short, this is a view that links fairy experiences of the past with present-day UFO events and begins to explain why countries like the USA and UK seem to experience a certain kind of abduction whilst other cultures, like those of sub-Saharan Africa, report very different experiences and far fewer UFOs.

This view puts UFOs and alien abductions into the realm of study that includes paranoid ideas and panics of all kinds. For example, Asia's koro epidemics, in which men become convinced that they are afflicted by a contagious disease causing their penises to shrink. Koro resembles UFO events in that reports come from frighteningly sincere individuals and have been documented to lead to occasional widespread outbreaks of general belief and mass reporting. The concentration on Asia would suggest a social dimension but there are cases around the world. In his Pocket Essential *Conspiracy Theories* Robin Ramsay reported an encounter with a scary individual who believed himself to be a robot whose every action was controlled from outside. He also believed that some unspecified people had broken into his house and substituted his penis for a smaller one![20]

There is nothing approaching a definitive investigation into how and why people see UFOs, but the tight consistency in a few studies is good evidence for there being a strong psychological component behind some experiences. However, as with all strands of ufological thought, there are obvious limits. These UFO state-of-mind theories look useful when examining some reports and facts. They probably don't explain why there were burns on Stefan Michalak's chest. They don't begin to explain what exploded over Tunguska in 1908. Most importantly, you might reject these explanations if you were to wake tomorrow morning with a strange memory of some bedroom visitor.

So, bear this research in mind because it is useful and it is all we have to help us understand how and why these things may happen to people.

Amazing Tales

The debates and the different angles have been considered, and we know it is unlikely that there will be any kind of agreement across the UFO community because each theory and each case often means what each community member decides it means. It's time to examine a few cases. This chapter has some of ufology's greatest hits and the next examines some strange and perplexing events. Armed with the insights gained so far, the conclusions are yours.

Roswell

If ufology has its own *Citizen Kane* or *Pet Sounds*, something so epoch-making that everyone is obliged to have an opinion on it, that case is Roswell. We can safely say that some event occurred in the summer of 1947 and by early July its significance was obvious to the highest-ranking elements of the US Army Air Force (USAAF) stationed at Roswell. Deemed insignificant for years, even by UFO investigators, the case took on monumental importance after its rediscovery in the late 1970s. Now the claims surrounding the event are pivotal to the standing of ufology.

It is claimed that on or around 1 July 1947 unknown targets appeared on the radar screens at White Sands Proving Grounds, a top-secret Army Air Force facility near Roswell, New Mexico. In the following days at least one local couple saw a glowing object pass overhead and rancher WW 'Mac' Brazel heard a loud explosion during a violent thunderstorm. Around the same time, he discovered a debris field made up of lightweight materials including printed shreds covered, apparently, in floral designs, 'I' shaped beams of balsa-like wood, tough metal foil and some 'hieroglyphics'. He took the material to neighbours Floyd and Loretta Proctor who suggested he turn it in to Sheriff George Wilcox. The sheriff notified Roswell Army Air Field (RAAF) Colonel William Blanchard who, in turn, despatched intelligence officer Major Jesse Marcel to interview the rancher and inspect the debris. Brazel, Marcel and counter-intelligence operative Captain Sheridan Cavitt headed out to the debris field, spending a day collecting material. At the Roswell base Colonel Blanchard decided to have the wreckage shipped to Wright Field in Ohio. He also oversaw a press announcement to the local media which stated that RAAF had in their possession a 'flying disc'. Predictably, this caused an instant sensation. As the debris stopped over at Fort Worth on the way to Ohio, General Roger Ramey took charge of affairs, ordering a press release stating that the material was nothing more significant than a crashed weather balloon. Wreckage consistent with this claim was displayed by Major Marcel in front of the press.

At this point the case went quiet for over 30 years, after which claims and counter-claims would cloud the real facts. Today some claim that Roswell was the site of the crash of an alien spacecraft, with variants of this story suggesting that at least one live alien was recovered. The number of recovered aliens, and the issue of whether they were all dead, varies with different versions of the story. Circumstantial evidence and a series of supporting witnesses support these pro-extraterrestrial claims. An alternative argument points to inconsistencies in the witness testimony and the conspicuous failure of the hard evidence produced to date to support this fantastic story. This generally sceptical camp see Roswell as a secret military operation that misfired, basing their argument on several key facts. It is an uncanny coincidence that aliens just happened to crash near the only fully operational nuclear bomber base on Earth, and were almost within sight of a key testing area for top-secret military equipment. The one strand of evidence that is used in both the pro-UFO and sceptical cases is a series of changes in the official explanations. These indicate that military authorities have been inclined to cover up and mislead. Beyond that, what these changes in the official line prove is debatable.

One of the most ironic points about this case is that the most vivid and widely disseminated versions of the story now disagree with the facts that started the whole craze. The generally accepted stories place the date of the UFO crash in early July but the one interview on record with Mac Brazel makes it clear that he found the debris in mid-June. He simply took it to the local sheriff on his next visit to town, around the Independence Day weekend. Similarly, there is doubt over the testimony of almost every key witness. Jesse Marcel, whose interview with ufologist Stanton Friedman led to the revival of the story, overstated his qualifications and experience whilst giving interviews to UFO investigators. He was widely ridiculed in the USAAF around the time of the Roswell incident for his own belief that the event was a flying saucer crash.

Another witness, local mortician Glenn Dennis, told a story that appeared convincing because it didn't present him as central or heroic. He claimed he had received phone calls seeking information about preserving bodies exposed to the elements for some time and requesting small sealed caskets. Subsequently, he had to drive an injured airman to the Roswell base. After delivering his human cargo he went in search of a girlfriend who worked as a nurse on the base. At the base hospital he realised things were out of the ordinary because of the presence of military police. His girlfriend emerged, clearly distressed and seemed horrified to see Dennis, warning him to get away because he could 'get killed' for sticking around. The pair, Glenn Dennis claimed, met for a drink the following week at which point the nurse told him she had been assisting at the autopsy of a dead alien. Dennis named the nurse as Naomi Selff, claiming she had been killed in an air crash following a British posting. Her whereabouts and real identity became a brief fixation in ufology. Initially, writers like Donald Schmitt claimed that the records of nurses at the Roswell base were unobtainable. By 1995, journalist Pete McCarthy had located the records, easily as it transpired, and tracked down the one surviving nurse, Rosemary McManus. McCarthy reported that McManus knew nothing of any retrieval of alien bodies and had never heard of Dennis's girlfriend Naomi Selff. Glenn Dennis (who died in 2015) was also a founder and stakeholder in the Roswell International UFO Museum and Research Centre.

Apart from Jesse Marcel there are a handful of others who claim direct involvement in retrieving debris or discovering the crash. All claim some variation on a story that suggests more than one crash site with a smattering of debris falling onto the ranch where Mac Brazel discovered it. In these accounts there were one or two other crash sites on which an entire craft, or virtually entire craft, hit the ground. The odd thing about these witnesses – Gerald Anderson, Jim Ragsdale, Frank J Kaufman and Grady L 'Barney' Barnett – is that none reported seeing the others at the site. Their testimony also differs in significant aspects of the case. Barnett's story was told by his friends Mr and Mrs Vernon Malthais because Barnett himself was dead. The disagreement in the stories has been taken by some to prove the worthlessness of the case. This is a little simplistic. Beyond doubt some of these witnesses are lying. All, however, introduced key elements to the story that have found at least one supporter.

Barney Barnett's account was the first to mention the presence of an archaeological team, working in the area, who had happened upon the crash site. Gerald Anderson was instrumental in supporting the surviving alien claim. Frank J Kaufman's story is incredible, placing him at the centre of events from the initial radar tracking, through the recovery of the craft, to the highly classified cover-up operation. Some of the more charitable researchers into the case claim that the length of time that has passed is bound to confuse memories. Maybe. But the fact remains that, to date, the Roswell case is saddled with an incredible story at its heart, key witnesses who disagree, a flexible number of crash sites and absolute confusion over the definitive date of the event. One key witness, Jim Ragsdale, moved the crash site in his testimony after a local UFO museum was refused permission to buy the original location he had identified. It is easy to be cynical but Ragsdale made his revised statement when he knew he was terminally ill. His video and book *The Jim Ragsdale Story*, in which the crash site was a crucial aspect, were intended to benefit his grandchildren.

We may dismiss much of the witness testimony but Roswell remains a live case because some witnesses are harder to put down and the official explanations are contradictory. Two military witnesses should be considered. Oliver 'Pappy' Henderson was a noted practical joker and his alleged UFO fragment shown to a fishing buddy was, in all probability, from a crashed V2 rocket.

Pappy, however, confided in his wife that he had flown top-secret wreckage to Wright Field in 1947. It is beyond doubt that such a flight did take place and went unreported at the press conference at which Marcel and wreckage which was clearly the remains of an ordinary weather balloon were posed for photographers. Henderson was one of the USAAF's top pilots with a history of involvement in key missions. He would have been a likely candidate for such an operation.

In 1990, Brigadier General Arthur E Exon approached researchers Kevin Randle and Donald Schmitt with the story they wanted to hear. Highly decorated, with a service history commanding Wright Field (renamed Wright Patterson Air Force Base in the 1960s), Exon, apparently, confirmed everything. Varied crash sites, recovery of bodies, confirmation of incredible properties of the recovered material and, the clincher, that the resulting investigation had established alien origin. Exon's standing appeared to put the matter beyond doubt. Randle and Schmitt's books *UFO Crash at Roswell* (1991) and *The Truth About the UFO Crash at Roswell* (1994) are consistent with Exon's story. But some key points should be noted. Exon made it clear that he had heard his stories as rumours and his timings are a little vague in the areas of key events, making it possible that he heard these rumours years after the alleged events. Exon himself wrote to Randle and Schmitt complaining they represented his account as fact when he had made clear its origin as rumour. Exon is disregarded by some commentators who don't trust his story. Others think his testimony may be intended to provide a case study in the credulous nature of UFO reporting. If so, the fact that those who quote him often treat his story as fact rather than rumour, has probably proven the point.

Exon's story is probably less fantastic than some of the official explanations to have emerged over the years. The United States Air Force (USAF) published *The Roswell Report: Fact Vs Fiction in the New Mexico Desert* in 1993, the same year that an investigation by the General Accounting Office (GAO) was started. The GAO is, in effect, America's Ombudsman's Department. The USAF settled on a combination of a misidentified Project MOGUL balloon and subsequent hysteria as their explanation. Project MOGUL was a top-secret undertaking in which weather balloons were used to carry sensors (particularly microphones), allowing them to pick up shockwaves from a great distance and monitor Russian nuclear activity. The 1995 GAO report established that most information management had been handled correctly. The report was in response to a Freedom of Information Act request filed by UFO researcher Karl Pflock. The GAO was not charged with discovering the truth of the event. They did discover that the pattern of communications traffic centred on the event showed no indication of a truly significant occurrence sufficient to demand attention at the highest level. UFO investigators seized on the discovery that the administrative records for the Roswell base between 1945 and 1949 appeared to have been destroyed. In 1997, the USAF followed their first investigation with a wider study, *The Roswell Report: Case Closed*. In a typical UFO community twist this report gave new life and heart to the very investigators it sought to silence. The new report attempted to explain accounts of military cordons and bodies in the witness testimony, concluding that mishaps in experiments with human dummies strapped to high altitude balloons had been the basis for these reports. The experiments and mishaps did occur but not until years after the alleged Roswell crash. The US Air Force said mistakes in witness memory in the intervening period had linked the dummy crashes with Roswell stories. Whatever the intention, the second Air Force report was a welcome development for the pro-extraterrestrial contingent whose case was in danger of being buried by contradictions by 1997. The loopholes in the official explanation convinced many that a desperate cover-up was still in place. Another view of this was succinctly stated by Lynn Picknett: 'The question is whether the USAF investigation set out to explain things that didn't need to be explained in the first place.'

Ironically the most damage to Roswell as the flagship case for ufology was probably done by those who most wanted to keep the case strong. With a grant from the Fund for UFO Research[1] Karl Pflock set about a comprehensive investigation of the case and produced *Roswell: Inconvenient Facts and the Will to Believe* (2001).[2] The subtitle of which suggests why there are so many anomalies in the story above. The inside covers of Pflock's book contain a map with nine locations marked, four of which are the alleged Roswell crash site including the 'conventional wisdom' crash site and a 'revisionist' crash site arrived at when one competing claim was reconsidered and joined the argument for the crash site put forward by other alleged witnesses. If you've struggled to follow the twists and turns of this short entry then Pflock's comprehensive demolition of the case might prove a 331-page slog and you could just content yourself with the opinions of John Keel – a revered figure in ufology – who called the Roswell case 'A boil on the ass of ufology'. You could also take on board that Pflock's use of an incredibly generous budget by the standards of UFO research took him far enough to conclude that the Roswell case started with the crash of a Project Mogul balloon – specifically NYU Flight 4 which launched on 4 June 1947. Predictably, the case remains alive and the most popular material remains the stuff that presents it as ufology hoped to find it: dead aliens, cover-ups and all. Season two of *UFO Hunters*, filmed years after Pflock's book, offered 'The Real Roswell', complete with an exploration of the desert in search of evidence. The rise of the internet has been a mixed blessing for ufology in general and the Roswell case in particular and this is probably the best point in this book to take a diversion to Wikipedia. In the early days of its existence Wikipedia excited many in the UFO community. Its open invitation to anyone to add to the body of knowledge convinced them that UFO cases would reveal themselves as credible and evidence-based. Wikipedia's editorial standards of verifying sources and supporting claims mimic the way peer-reviewed academic articles have to stand up their evidence, in effect creating a 'lite' version of the same process. Many of the most valued UFO cases, Roswell included, have been shredded online as a result as Wikipedia's terse prose and demand for sources reduces once florid arguments to a list of inconsistencies and observations that someone 'reported' or 'claimed' a thing to be true. The most vivid moments in those stories most cherished by the UFO community sometimes appear on Wikipedia next to the ubiquitous 'citation needed'.

Roswell isn't completely dead in the water, as its regular appearance in television documentaries and popular books on UFOs proves. But anyone taking in the published history of the case now is faced with isolating the credible witnesses in a pack of liars and assuming that the case continues to be classified at the highest level because somewhere in the morass of disagreement and disharmony there is a truth so monumental that, to date, the few who genuinely know it have deemed it too explosive to share with the rest of the world.

You could, of course, turn back the clock and enjoy the breaking excitement as it was around the turn of the century by reading those books from the 1990s that rode the wave of emerging witnesses and growing excitement that the truth about a crashed UFO was about to be revealed.[3] This is not a facetious suggestion. We'll consider the value of reading the UFO literature of long ago later in the proceedings. Roswell – the town – continues to prosper on tourist dollars and with increasing scepticism about the case that made the town famous has come some creativity and adaptation to the new realities. Should you find yourself in the neighbourhood you could always visit the International UFO Museum and Research Centre which 'endeavors to be the leading information source in history, science and research about UFO events worldwide'.

Murmurs in the Ministry

According to former MOD man Ralph Noyes the following cases were amongst the handful that had his employers in a genuine state of anxiety. Military cases, 20 years apart, both involve a complex series of events with a range of evidence.

At 9.30 pm on 13 August 1956 radar operators at RAF Bentwaters picked up a target over the North Sea moving towards the base at incredible speed. Estimates would later place the velocity around 4950 mph. Nothing man-made could move that fast in 1956. When this object vanished from the scope, eighteen other targets, led by a group of three, appeared. Once again moving inland from the sea, the second wave were on a different trajectory and moving more slowly. A T-33 aircraft, vectored to investigate, saw nothing. The cluster of targets then appeared to group together to form one target, giving a stronger radar echo than any known aircraft. The radar returns convinced the radar staff they were dealing with something real.

The large object was followed on radar until it vanished. Soon afterwards a target moving at the same incredible speed as the first was detected. Just before 11 pm the staff in the radar tower saw an object out of the windows. The presence of something strange in the sky was also confirmed by the pilot of a C-47 transport aircraft who reported a bright light streaking underneath his aircraft. Within minutes an open line had been established between radar operation rooms covering Norfolk and Suffolk at Bentwaters, Lakenheath and Sculthorpe. Lakenheath located and tracked a target that alternated stationary periods with movement between 400 and 600 mph. Training Sergeant Forrest D Perkins at Lakenheath Air Traffic Control was coordinating the radio traffic and was able to establish that a separate radar operation monitoring ground approach in the same area also had the object on their scopes. This tracking alone struck him as bizarre given the presence of a Moving Target Indicator (MTI) on the radar equipment. This should have disregarded stationary objects. One possible explanation for the continued tracking of a stationary object is that the whole object is rotating or showing rapid movement on its surface.

An air intercept was ordered and a Venom aircraft from RAF Waterbeach near Cambridge was scrambled. By midnight the pilot had visual contact and his radar operator had a lock on the object. The pilot reported, 'I've got my guns on him.' Within seconds the pilot had lost sight and his radar operator was reporting that the object had moved rapidly behind them. Confused and quite possibly frightened, the pilot called for instructions but the operators manning the military scopes had never seen anything like the movement of the object. Trained to monitor combat, they recognised the development as potentially hostile but couldn't advise on a strategy for the pilot. The obvious response, scrambling a second Venom, was effected. Within a short time the object had vanished from all but one of the scopes. This development is also significant. The object hadn't vanished. In all probability it had just been seen by civilians on the ground with the Venom in pursuit. Only Lakenheath's Air Traffic Control radar had the capability of monitoring low level manoeuvres. They followed the object for a further five minutes as it pursued the first Venom and then stopped at very low altitude once again, before disappearing from the scope for good. The second Venom, arriving on the scene later, got no visual or radar lock.[4]

Just over 20 years later a similar event shook the UK and US military, and the Iranian Air Force who experienced it first-hand. At 12.30 am on 19 September 1976, the occupants of the Shemiran area of Tehran reported a brilliant light overhead. Initially sceptical of the reports, Deputy Commander of Operations BG Yousefi received a report from the local Mehrabad Air Traffic Control that they had a track of the object. Yousefi went outside and saw the object for himself. By 1.30 he had ordered an F-4 Phantom from Shahrokhi Air Force Base to intercept. Twenty-five miles from the object the pilot had visual and radar confirmation of his target. Having confirmed a radar lock he was stunned to see all his instrumentation, along with the UHF and VHF radio, fail at the same moment. Helpless to do anything other than fly the plane, he headed for base. All his suspect systems returned to full working order halfway back to the runway.

Another F-4 was already on its way. The second aircraft, piloted by Lieutenant Jafari, got a lock around 27 miles. Jafari and the UFO were both travelling at speed but his instruments indicated he was gaining on the object at 150 mph. The fighter was travelling well above this speed, indicating the object was moving quickly away from the aircraft. The radar return on his scope indicated the object to be around the same size as a Boeing 707 tanker aircraft. Soon the object shot away. Jafari, breaking the sound barrier on full afterburner, tried to close the distance. Then things took a turn for the worse.

A small object detached itself from the large target and flew directly towards the F-4. Jafari armed an air-to-air AIM 9 (Sidewinder) missile to respond to this apparent rocket attack. Within seconds his plane, like the other F-4, was in the throes of a full electronic systems failure with the unwelcome complication of an imminent impact. Jafari took desperate evasive action, hurling the fighter into a rapid turn and dive manoeuvre. The small object followed the Phantom at a distance of 3-4 miles before returning towards the larger UFO. Whatever this small object had been it certainly wasn't the simple air-to-air missile Jafari feared. The Phantom's systems returned but Jafari wasn't about to attempt another attack.

The large UFO released a second object which headed down into the desert. Jafari and his navigator watched, expecting a missile explosion. They saw the new object appear to soft-land on the ground before bathing the desert with bright light. Homing in on the location of the landing, Jafari noticed pulsing electrical interference, possibly from the object. Fifty miles away an airliner coming into Tehran reported a similar problem identifying the source along the same compass bearing as reported by Jafari.

Whatever the cause, the brilliant glow of the object had damaged Jafari's night vision to the point that Shahrokhi had to guide him back in. He needed several circuits over the base before he was ready to land. The following day the crew of the second F-4 were helicoptered to the desert location. The occupants of the only house for miles reported hearing a loud noise and seeing a brilliant light the previous night.[5]

In an utterly incredible twist to this case, researchers Jenny Randles and Peter Hough talked to Dr Simon Taylor, a British university lecturer, then engaged and later married to an Iranian woman. Camping out on a mountain top near Tehran just hours before the mid-air incident he had experienced an event almost beyond description. Phrases like 'altered state of consciousness' barely do justice to an incident in which the lecturer and a friend found themselves open to surreal experiences that bordered on dreams, leaving them with short-term problems of adjustment and some longer-term health problems.[6]

To date, no widely accepted explanation is available for the Bentwaters or Tehran cases. Other similar tales exist but these two are, arguably, the best, given the range of evidence from radar to eyewitness, the length of the sightings, the number of witnesses and the proven presence in both events of supporting documentation from military sources. As described above, these cases hint at a phenomenon which appears to display intelligence and a direct ability to respond to unwanted human investigations. Regardless of the cause, these were events that caused concern in military circles, not least because, in their initial stages, they resembled hostile enemy activity.

That final point, however, may be the main reason the British military displayed the concern discussed by Ralph Noyes and it should be noted that sceptical opinions have been applied to both cases. The details that gave the Bentwaters case of 1956 such power in a host of published works about UFOs rely very much on the fact that Forrest D Perkins – who was on duty that night – presented the story as described above to the Condon Committee (one of a series of investigations the US government established into UFOs). They also rely on press coverage of the case in the UK in 1978 which prompted responses including one from Flight

Lieutenant Freddie Wimbledon who had been on radar duty that night and involved in the incident. He disagreed with Perkins about who had directed the airborne intercept (suggesting it was his British team whilst the Americans were listening in) but agreed with the rest of the story. By contrast, Jenny Randles, Andy Roberts, David Clark and Martin Shough conducted interviews with witnesses in the late 1980s, turning up air crew in and around the incident who disputed key details, notably the moment at which the target had suddenly located itself behind the Venom aircraft.[7] Sceptics on the Tehran incident start by breaking down the various anomalies into individual problems at which point the apparently impressive series of events looks less spectacular. Aviation journalist Philip Klass was for many years the go-to sceptic for such interpretations and his book *UFOs: The Public Deceived* rips into the most remarkable claims, suggesting there was electrical malfunction on only one aircraft, pointing out that this mission was the first ever night flight for the aircrew involved and – crucially – presenting an image of the Iranian Air Force in 1976 as something of a rich boys' club with little substantial combat readiness. Klass also suggested the supposed mother ship was a misidentification of Jupiter. Given that the fighters took off heading for the north of the city the mothership was exactly where Jupiter would have been in the sky at that time.[8]

Bass, How Low Can You Go?

Another worrying case occurred on 21 October 1978 in Australia. 20-year-old Frederick Valentich, an amateur pilot who dreamed of flying professionally, had hired a Cessna 182 for a short return flight from Moorabbin Airfield near Melbourne to King Island. His intention, apparently, was to pick up crayfish. He filed a flight plan but an unexplained trip from the airfield rendered this useless and ensured he would arrive at King Island after dark. Once airborne the young pilot began to ask Air Traffic Control (ATC) about unknown aircraft in his area. Flying below 5,000 feet, he was off the radar scope. So was the bright green light he described flying over and above him. The radio recording of Valentich's conversations with Steve Robey at ATC is one of ufology's most puzzling pieces of evidence. As the event escalates and Valentich reports his engine rough idling, other comments are made. 'It's not an aircraft, it's…' Valentich never completed that sentence. He had remained calm but his final communications to ATC betrayed concern: 'That strange aircraft is hovering on top of me again… It is hovering and it is not an aircraft… Delta Sierra Juliet, Melbourne…' A burst of noise, variously identified as carrier wave interference and 'metallic scraping' closed the communication for good.

In May 1982, the official Australian investigation concluded, 'The reason for the disappearance… has not been determined.' There has been a lot of speculation in the UFO community. Sceptics point out that Valentich had an active interest in UFOs, his disappearance from the airfield ahead of the flight was never explained and he took much more fuel than he needed. In addition, he was an inexperienced pilot attempting a difficult night landing and had failed to carry out a standard procedure, calling ahead to King Island to ensure the runway was illuminated. In short, these facts might suggest he was planning a major UFO hoax. Those closest to the pilot – notably his father – reject this. Valentich still had time to radio ahead and apparently spent much of his short flight bemused and then frightened by the strange light. One intriguing theory, advanced by Dr Richard Haines, suggests that Valentich might have fallen foul of a covert military experiment which generated the light phenomena. There is some circumstantial evidence for this in a series of events in places like Orford Ness on the East Anglian coast where light phenomena, experiments in Star Wars technology, and some unexplained air accidents have occurred in close proximity.

A more prosaic explanation has been offered by Brian Dunning in a *Skeptoid* episode. He cites evidence from Valentich's fledgling aviation career to show that in earlier flights he had already committed a few transgressions both by making an incursion into restricted airspace and by flying into a cloud deliberately. Dunning goes on to point out that the final radio message is uncannily similar to one featured in a scene from *Close Encounters of the Third Kind*, suggesting in effect that Valentich preplanned this part of the flight but then found himself dangerously disoriented and over water. Dunning suggests: 'He circled to give the radar guys something to see, possibly starving his carburetors. Suddenly he was paying attention to his engine instead of to the horizon, in the dark over water for the first time; and before he knew it he had 178 seconds to live.' The '178 seconds' being the available recovery time known to air crash investigators from their study of a number of similar incidents.[9] A substantial piece of wreckage, almost certainly from Valentich's plane, was subsequently found in Tasmania. The depth and inaccessibility of the wreck have probably kept it from the search teams. Whether the discovery of the wreckage would bring us any closer to an understanding of this chilling case is debatable.[10]

Aldershot Alfred Abducted

As we have seen, the debate on abductions is fierce. Admit hypnotic regression into the argument and you have some researchers claiming millions of abductees worldwide. Other researchers, along with professionals in areas like psychotherapy, now accept the dangers of hypnosis. So, forget every case that involved hypnosis and forget those in which the witness may conceivably have been led by a researcher with his or her agenda, and you eliminate many abductions. Now rule out any remaining abductee who gained notable status, fame or money through their experience. This reduces us to a basic hard core of cases. But they do exist. As I mentioned earlier, I met one experiencer whose fear of notoriety led him to demand that his wife never found out. I have no idea what had really happened to him. But the following case does count as an abduction, and it still stands if we rule out everything listed in this paragraph.

In the early hours of the morning on 12 August 1983, 77-year-old Alfred Burtoo was fishing beside the Basingstoke Canal in Aldershot. He saw a light approaching in the sky and noticed that it dropped out of sight some distance from him. Soon two beings, humanoid in appearance, around four feet tall and dressed in coveralls, appeared and beckoned him to follow them. They went to a landed object that Burtoo would later sketch. Broadly similar to the typical saucer designs, the object did have odd ski-type legs. Burtoo described the object as black inside and like burnished aluminium on the outside. A voice, from a source he couldn't see, spoke to him. 'What is your age?' it asked. Burtoo pointed out he would be 78 on his next birthday. 'You are too old and infirm for our purpose,' replied the voice. Soon afterwards Burtoo departed, walked back to his seat on the canal bank and was happily chasing fish when the object departed with a bright glow. The glow was so bright it allowed Burtoo to see his float bobbing near the far bank of the canal. Burtoo had lived a tough and eventful life. He'd seen military service including active duties in World War Two. He'd also trapped and hunted in Canada, confronting the danger of bears and wolves. He suffered no obvious physical effects from his UFO encounter but he did admit that turning the events over in his mind caused him problems in getting to sleep. He died aged 80 having consistently told the same story to family, friends and the odd UFO researcher.

If Alfred Burtoo was lying he went against the habits of a lifetime, misled a wife with whom he'd enjoyed a long and happy marriage and deceived the many people who respected him as a local historian. In return his greatest achievement was fleeting fame in a local Aldershot paper – achieved not because he wanted fame but because he approached them in the hope of locating another witness to the UFO. His story appeared in a million-selling book, but not until he was already dead. If he was telling the truth then he presents us with an apparent paradox. We have a craft and beings who possess a technology well ahead of our own and the ability to concoct a mission that seems both complex and important. They jeopardise the lot because their advanced abilities still leave them unable to tell a pensioner with a fishing line apart from younger and fitter specimens of humanity.[11, 12, 13]

Wow!

At one point in the Abduction Study Conference held at Massachusetts Institute of Technology in June 1992 there was an animated argument, the nub of which involved UFO researchers berating SETI (Search for Extra-Terrestrial Intelligence) researchers. The argument from the UFO crowd was that SETI had spent tens of millions of dollars producing nothing whilst UFO investigation was dealing directly with witnesses and inexplicable events and surviving mainly on donations and volunteers. The argument got personal, and the two sides arguing that day have often exchanged unpleasantries. Ironically, both sides say they seek the same thing. Ironically too, if you got representatives of either side alone for long enough, they might admit they agree that the closest they have come to their goal is the event of 15 August 1977. On that day a strong narrowband radio signal was received by the radio telescope known as Big Ear and operated by Ohio State University. The signal – apparently from the constellation Sagittarius – appeared to be extraterrestrial in origin. Everyone at Ohio State went home that night without worrying themselves about it because it was not until a few days later that astronomer Jerry R Ehman reviewed the computer printout. His circling of the computer reading along with the scribble he put next to it – Wow! – has given the event its name. The technicalities are easy to prove, the signal was observed for 72 seconds before the radio telescope searched another area of the sky. The signal had no detectable modulation. The exact frequency remains debatable but only in a very narrow range. It was definitely near 1420.4000 megahertz with a bandwidth of less than 10 kHz and the exact location of the signal is impossible to determine because it was only picked up on one of the radio telescope's two horns. The operating capabilities of the equipment are such that it is impossible to determine which horn picked up the signal, making it hard to pinpoint exactly where in the constellation Sagittarius it came from. If indeed it was from there and was an intelligent message.

The Wow! signal matters because, since the moment Jerry R Ehman scribbled his response next to the printed readout, the best minds in SETI have applied themselves to making sense of the signal. The obvious actions of searching the same piece of sky with each generation of radio telescopes have produced nothing. For a while Ehman considered it possible he'd spotted an earthbound signal randomly reflected off passing space hardware but was gradually persuaded this was highly unlikely. The problem with all scientific theories advanced to explain the signal to date is that they are all unlikely. But the Wow! signal was real. Ironically, it could be gifting SETI the very thing it is seeking but – if so – it's also a scientific nightmare because 'Isolated events are not science's strong suite. If something happens once and never appears again then it's difficult, if not impossible, for the methods of science to be of much use.'[14] By contrast this is not the kind of proof most in the UFO community were hoping would form the best evidence of extraterrestrial life. But the tantalising enigma remains. Ehman has considered the origin may have been military, the implication being those responsible may have their own reasons for remaining quiet. But – truthfully – nobody knows. SETI and ufology, often at odds with each other, can't let this mystery go and, in this example, if nowhere else in this book, everyone is on the same side in wanting an answer. The answers they imagine may, however, vary a great deal. But where the Wow! signal is concerned everyone behaves a little differently. Brian Dunning's *Skeptoid* demolitions of ufology's worst moments have been unsparing but in the case of the Wow! signal he stated: 'In conclusion, yes, an alien intelligence is still a candidate explanation for the Wow! signal. But there's no evidence for this. A stronger candidate is the significantly more vague explanation of an interstellar radio source of unknown origin.'[15] Who knows? The signal's one-off appearance could be the giveaway. Perhaps the space brothers were calling back one of their own. Elvis 'died' the following day.

All of the above – of course – concentrates on those spectacular stories involving claims of contact with intelligent alien life. Much scientific opinion suggests the likely first proof of alien life will involve simple life. Since the discovery of life-forms known as extremophiles (organisms attuned to conditions most life forms would find abhorrent) on Earth, speculation has increased that similar creatures may exist throughout space. The announcement in September 2020 of the apparent discovery of phosphine gas in the Venusian atmosphere currently provides the likeliest clue that we are not alone. Microbial life forms, living in that planet's atmosphere, an environment that would be fatally toxic to life on Earth, could be emitting the gas. As *Science News* expressed the evidence at the time: 'If the discovery holds up, and if no other explanations for the gas are found, then the hellish planet next door could be the first to yield signs of extraterrestrial life – though those are very big ifs.' Big ifs indeed, scientists at the time were generally happy to conclude the finding to be a 'possible' sign of life. Significantly, there were no other obvious explanations for the evidence gathered. If there are microbes in the atmosphere of the second stone from the Sun we can be confident they are not responsible for the Wow! signal. Where the strange object named Oumuamua (a Hawaiian word meaning "first scout from a distant place") is concerned we can't be so sure. Oumuamua passed through our solar system, coming closest to Earth in late 2017. Harvard astronomer Avi Loeb is amongst those intrigued by the observations made and willing to consider the evidence for the object being the construction of intelligent beings. His book Extraterrestrial: The First Sign of Intelligent Life Beyond Earth (2021) tackles the evidence directly. It's also a good place to go next, if the arguments in this chapter have grabbed you.

Cautionary Tales

Ufology is partly infotainment – information gathered into market-friendly packages. The bestselling books on ufology deal with abductions and the belief that UFOs are alien spacecraft. This is possible but not proven. Television loves a case with photo and video evidence – often to the point they can present this material and ignore the fact it was explained away in rational terms years before. Particular favourites in this area include the Belgian UFO wave (loads of inconclusive sightings and one good photo for which the hoaxers have long claimed credit and provided evidence) and the Phoenix lights (decent footage, lots of good eyewitness accounts filmed but pretty much an open secret that most of the fuss in this case started when five A-10 aircraft took part in a flare drop exercise miles away from Phoenix).[1] So, we should be cautious when the answers appear simple. As we have seen already in this book, the answers to UFO mysteries may come from some unlikely areas. Here are some examples.

'The Wings and Engine of a Flying Saucer.'

On 21 March 1953 a top-secret shipment arrived in New York aboard the USS *General AW Greely*. Two days later, the cargo was safely deposited in Fort Knox, Kentucky, where it would remain for a quarter of a century. Prior to the sailing of the USS *General AW Greely* from Bremerhaven in Germany, the cargo had been under top-secret American guard in Austria. The men guarding the cargo as it moved were told they were transporting 'The wings and engine of a flying saucer'. This much is undisputed fact because it is confirmed by a declassified memo from the US State Department. The memo has been declassified because there is no longer any need for secrecy in this case.

The boxes contained the Hungarian crown jewels, including the Crown of St Stephen. This one relic alone is now officially priceless. When it did have an estimable value the price was an astronomical $300 million, and that was in the 1930s. Spirited away from Hungary by patriots who didn't want the relics to fall into Communist hands, the treasures were eventually returned by US President Jimmy Carter in 1978. Today they are on public view in Budapest.

So, decades ago, the American State Department lied to its own servicemen, albeit for reasons of security. These lies were UFO stories and today the stories resemble several widely believed UFO tales. Some crash retrieval stories have identified Fort Knox as the location of extraterrestrial wreckage. But the highly fortified bullion depository is an impractical place to carry out any kind of investigation on such material. Other elements of the story – the top-secret movement of sealed cargo, the minimal briefings given to those guarding the material and the need for a highly secure destination – are mainstays of the best-known crash retrieval stories.

We know this story is true because the return of the treasures to a politically secure homeland has released classified documentation. We don't know how influential the State Department's cover story became in the development of the crash retrieval legends that fill books and websites today. There are other stories and other servicemen who swear they moved alien artefacts to top-secret locations. Some of these people may be after publicity but others are certainly telling the truth as they know it. Or, more accurately perhaps, the truth as they were told it.[2]

'Can't You See Him?'

On or around 22 February 1973, one of the strangest and potentially most significant abduction events in UFO history occurred. Whilst the exact date of this event has been lost in claim and counter-claim there is no doubt about the start of the abductee's involvement in UFO experiences. Maureen Puddy's first UFO encounter occurred on 3 July 1972 around 9.15 pm as she drove to visit her son in hospital just south of Melbourne, Australia. Seeing the road bathed in blue light she initially thought the source was a helicopter, similar to the one in which her son had been carried. When the light appeared to be following her, she stopped, got out and saw a large craft which caused her to panic. An eight-mile chase ensued which Maureen Puddy reported to the police and the Royal Australian Air Force. She later had other experiences including further sightings and telepathic contact with a being aboard a UFO. In late February 1973, she had arranged to meet UFO investigators Judith Magee and Paul Norman at a remote spot where, previously, she had experienced her car stopping during a strange encounter. On the way to the meeting Puddy saw a spaceman in a golden suit appear and disappear from her car.

Puddy was reporting this experience to Magee and Norman

at their designated meeting spot when her spaceman appeared again. 'There he is. Can't you see him?' said Puddy, but the two investigators saw nothing. Things got stranger. The two investigators were sitting in the back of Puddy's car as she described her visitor moving outside. Norman got out of the car and walked round. Puddy claimed the spaceman had moved to allow Norman to pass. Then, with the two UFO researchers in the back seat of her car, Puddy appeared to experience a full-blown abduction. Seemingly conscious at first, she later went into a faint but still managed to relay details of an event that we would recognise today as a fairly normal abduction. She claimed to be aboard the spaceman's craft but she remained firmly visible in the driving seat to Magee and Norman.

Maureen Puddy went through a range of emotions, ending up in tears at one point as she was told by her hosts aboard the UFO that she would remember nothing of the incident. The two researchers remembered and recorded everything, although the exact date has since confused them! Puddy had other experiences, receiving mainly ridicule as a result. Like many abductees she felt confused and bewildered by the way the phenomena appeared to have chosen her. Her experiences predate the era when UFO abductions were a widely known event. In Australia in the early 1970s, public knowledge of UFO events was sketchy.

What exactly happened to a woman who described herself as a 'housewife' remains a mystery. Researchers like Albert Budden claim Puddy was someone suffering hypersensitive reactions to electromagnetic events. In this scenario we are dealing with hallucinations that can be brought on by events as simple as a car travelling a long distance and building up an electric charge. Budden's theories have a scientific basis and Puddy's case forms the final pages of his book *Electric UFOs*.[3] Another scenario might suggest that some external intelligence really did contact Maureen Puddy on several occasions, taking control of her senses and trying to communicate with her. If so, it is hard to say why this choice was made and what useful purpose was served.[4, 5]

'Cold!? Just Don't Ask.'

The Hessdalen Valley, south-east of Trondheim in Norway, is remote, thinly populated and little known. For years the valley has hosted sightings of strange lights. There is some doubt about when the phenomena started but, in 1981, the reports were regular and reliable and they continued long enough to allow a dedicated team of UFO investigators to establish a research project. Project Hessdalen went public in the summer of 1983. Active investigations started that year with a smattering of equipment and the project was fully equipped and running by 21 January 1984, by which point it had already gathered some incredible findings. The universities in Bergen and Oslo along with Norway's Defence Research Establishment provided equipment to monitor magnetic events, photographic gear including spectrum grating equipment designed to indicate the solidity of the objects, a radar tracking rig and a laser.

The events showed fluctuations with periods of high and low activity but the team gathered dozens of photographs and some other data that remain hard to explain. Within the first month they had two genuine revelations. The radar tracking equipment showed small objects moving frequently and the team were sometimes unable to spot these objects with the naked eye. On 31 January 1984, a lengthy series of radar recordings was made over a number of hours, all indicating objects within sight of the Project Hessdalen caravan. Repeated observations by the researchers found nothing visible in the air. This pattern was repeated over the whole investigation but, on many other occasions, the radar tracks matched with sightings made by researchers and local residents.

Nine times the team pointed a laser at the lights, noting a change in the behaviour of a target on eight of these occasions. The most spectacular such event happened around 7.30 pm on 12 January 1984 when a laser was pointed at a light that was flashing regularly. When the laser hit the light, it doubled the frequency of its flashes. When the team turned off the laser the light returned to the first sequence. The exercise was repeated four times, showing the same result each time.

Whatever the team were dealing with, it was a genuine mystery to science. Results from the spectrum gratings on the cameras indicated the objects photographed were solid. Radar traces indicated that these solid objects could move at supersonic speeds without creating a sonic boom. Put bluntly, these objects showed properties fitting a solid and non-solid. On many occasions the team observed lights that changed colour and brightness. In some cases, lights simply faded out completely whilst under observation.

Two Hessdalen radar images were forwarded to an American organisation, Ground Saucer Watch (GSW), specialising in expert photographic analysis of UFO cases. Their reports have been the graveyard of many hopeful cases. Their analysis indicated that the Hessdalen radar images appeared to be from a good radar-reflecting source which was, apparently, more dense in the centre. Perhaps the most insightful comment they could offer was that, 'The return appears to be more indicative of one from a water-laden cloud.'

Strangely, the phenomena appeared to be aware of the investigators. There were several equipment malfunctions, never properly explained. Some malfunctions continued after the equipment had been dismantled, investigated and checked for problems. On two consecutive nights lights appeared on radar within a minute of the videotape used to record the screen running to an end. On other occasions team members would arrive in their cars, be confronted with a sighting immediately and then go to the caravan for a night in which they would record nothing. One local resident once reported feeling the pressing urge to walk outside her house. When she did, she saw a light passing by. All of this alleged intelligence shown by the lights could, of course, be nothing but coincidence.[6]

The Project Hessdalen team produced hard scientific data in the most extreme of conditions. The findings may be expressed in terms of scientific terminology but the reality was that there were a handful of dedicated researchers with borrowed equipment. The centre of investigations was a lonely caravan in which the research decisions involved the grim realities of unplugging heaters to free sockets for radar equipment. We're talking Norway in January. I once asked Hessdalen founder member Odd-Gunnar Roed about the winter temperatures in the caravan. Considering he is one of the few ufologists in history who has been able to go to his investigations with real expectations of having his own experience Roed is remarkably laid-back. Ask him about those winter nights and he gets animated. 'Cold!?' he sighed letting a slow resigned shrug of his shoulders speak volumes, 'Just don't ask.' The project, and the lights, have kept a low profile – currently averaging around 20 sightings a year, mainly monitored now by automatic equipment. Project Hessdalen has produced real UFO evidence. But, evidence for what?[7]

The researchers believed they were dealing with a strange atmospheric phenomenon. A few sightings may well be misperceptions of planets or car headlights viewed under unusual atmospheric conditions but this doesn't explain the most mysterious events. There is no doubt the Hessdalen Valley hosts phenomena unknown elsewhere so the most promising theories take on board local conditions. One suggestion is that hydrogen, oxygen, and sodium flare up in combustion because of large scandium deposits in the area. A more complex idea suggests macroscopic Coulomb crystals cluster in a plasma. This plasma is likely to be the result of ionization of air and dust by alpha particles during radon decay in the dusty atmosphere. Truthfully, it's still a mystery but it does bring us back to the existence of earthlights and the work of researchers like Paul Devereux. It also reminds us that, in this case, we have UFOs we can't argue away as a hoax or misperception. Indeed, as long ago as 1986, rock crush experiments at the US Bureau of Mines produced minute luminous forms in the air.[8] It's been a belief of earthlight proponents for years that these experiments, imagined up to the scale of parts of the planet's crust rubbing up against each other could produce some of the phenomena described over the years as fairy lights or UFOs. Nobody is suggesting these lights at Hessdalen or elsewhere are alien in origin.

'You Bastard!'

I got a phone call at work one day from Justin Williams, a journalist on the local *Kent Messenger* who was researching material for a series called 'Kent's X-Files'. He'd heard me a few days earlier doing a local radio programme on UFOs and wanted to check out any leads I might have. A few days later we were sitting either side of his desk at the *Kent Messenger*'s Larkfield base. One event he raised with me was the Dargle Cottage case. An obscure number by UFO standards, but local to Kent. The case revolved around an elderly couple. Mr and Mrs Anthony Verney lived in a semi-remote Kent cottage in the late 1970s and early 1980s and claimed their lives were ruined by strange phenomena from a secret bunker nearby. Assuming the whole thing wasn't a paranoid delusion, the likeliest explanation is that a military bunker was being used for experiments with electronic equipment, possibly weaponry, designed to direct low frequency radiation. Anthony Verney wrote copiously to complain and demand action. His correspondence with then Prime Minister Margaret Thatcher is one rare case in which a private individual outdid her for assertiveness. The case, though long dead in terms of activity, still makes the UFO literature on two counts. Firstly, the possibility that electronic mind-control experiments could cause UFO-like events supports the theories of people like Albert Budden. Secondly, it appeals to those with a belief in conspiracies because the terse official replies look for all the world like blanket denials. Dargle Cottage makes a fleeting appearance in Robin Ramsay's Pocket Essential *Conspiracy Theories* where the basis and context of the alleged experiments are explained.[9]

Driving home from Justin's office it struck me that I'd never written anything about the Dargle Cottage case myself. In that brief period when UFO magazines were littering the newsstands there would be money in an easy piece of writing. The following Saturday, 13 April 1996, with my wife away at her annual Psychotherapy Conference, I'd already promised to take our son Thom for a special treat. He was three and a half, loved Thomas the Tank Engine and wanted nothing more than a ride on the local Tenterden Steam Railway. We had a brilliant afternoon and Thom fell into a contented doze as we started the drive home. Dargle Cottage is located on a quiet road near that steam railway. I wanted a photo of the sign on the gate of the property, and the cottage isn't visible from the road. The site of the alleged experiments is underground on farm land, which by 1996 had been turned into a pitch-and-putt golf course. I could get pictures of both locations inside 15 minutes and still be home in good time.

I was just getting the second shot when a car came round a corner and pulled into the side of the road. When I drove away, it pulled out to follow me. The couple inside looked a little like Mulder and Scully. Young, smartly dressed and wearing shades. Not exactly normal for the Kent countryside on a Saturday afternoon. I drove to the main road, picked a lay-by and dived in at the last minute leaving them no time to follow me and no room to stop. I took their number, dropped back and followed them. A few miles later they were on my tail following a rapid reversal out of a farm road onto the main road. This was for real! My young son was fast asleep on the back seat and, if M15 were going to make me pay for stumbling onto a biggie, I didn't want him in the firing line. We got to Maidstone and I lost them, turning late into a small series of back roads near my home. Even if they'd followed me, I knew these roads well. I knew I had a chance to lose them. Mulder and Scully were nowhere in sight when I got home.

The following Monday I rang two people, a mate who could trace their car number and Justin Williams. 'I really want to speak to you,' he said, 'There's still something happening at Dargle Cottage.'

'Yeah, tell me about it,' I said, launching into my experience. I'd hardly started when he let rip.

'You bastard!' he shouted.

The rest fell into place in seconds. He'd been driving the other car with his colleague Beth Mullins beside him as they did their research for 'Kent's X-Files'. Three days before we'd been no more than six feet apart for about half an hour, both of us agreeing that we had an interest in the subject but had a sceptical and pragmatic approach. The previous Saturday afternoon neither of us had recognised the other one. We'd gone further than that. I'd seen two people who didn't belong in that area and drawn a blank trying to reason it away as anything but some security operation linked to Dargle Cottage.

Justin had even more evidence to scare him. He'd already run a check on my number plate and linked the car to the Kent County motor pool. A newish Rover saloon that belonged to the county. To him this meant one thing. As a journalist he knew that undercover operations used cars just like this, and from the same source because they didn't look out of place anywhere. He also noticed the two security passes in the windscreen of my car and did not realise they were one for my academic job and one for my wife's work at Kent County Council. He connected them with the place towards which I'd led him that afternoon. We'd come into Maidstone on a main road that brought us right past County Police Headquarters. The point at which I'd dived out into the country to lose them would have allowed me to do a simple diversion and quickly get back to Police Headquarters. Justin had seen a man with a camera standing next to a lonely gate and looking directly at a location he knew to have some links with an unexplained case. The same person got into a saloon car, spent time chasing him and Beth along two roads, and showed

119

every sign of wanting to get back to Police Headquarters. Justin was convinced he'd stumbled into some trouble, until I rang him.

For a couple of clear-thinking, intelligent sceptics we were a disgrace that afternoon. Now, I can see this as a lucky break. It brought several things home to me. For starters, the things we saw were the things we expected to see once it was obvious there was a chase on. I didn't see, or consider I might be looking at, Justin. Frankly, I wasn't thinking beyond getting Thom out of harm's way. Justin didn't recognise me or see the child seat in the back of my undercover police car. We were always going to sort it out because both of us desperately wanted to speak to the other the following Monday. But there must be many other cases where people don't make such connections.

Outside of the world's official UFO tourist attractions, like the assorted delights around Roswell, there are hundreds of unofficial locations. In the UK alone, people stumble to places as remote as Rendlesham Forest, a mountain near Bala in Wales and the forest clearing in which Bob Taylor's trousers came to grief. They stop, look around, prod the ground and generally do things that others might consider suspicious. They want that proximity to a place where a UFO incident is known to have occurred. Rationally, they don't expect to find a thing. But these pilgrimages are not about being rational. They are about wanting contact with a mystery that remains elusive. I'm not having a go. I'm one of these people! I've been to the places described in books and websites, looked and wondered. Sometimes, I've come to my own conclusions but I've never learned more than I did as a result of that April afternoon.

What I know from personal experience now is that your mindset at these moments is everything. Put yourself in a UFO-related location, suspect that the place might be under surveillance and you only need a couple of hydro-geologists in the area, dressed in coveralls and taking a soil sample to a Land Rover, to prove you were right about that UFO crash all along. Seriously, for a few minutes on that Saturday afternoon in 1996, the world became completely surreal. Even the breaking sports news on the radio had gone mad. When I needed to check the date to write the incident up it was easy because I'm a football fan. That date is effortlessly located because of what I was hearing as the car chase went on. As Mulder and Scully played tag with my car, Alex Ferguson was busy telling the BBC that his Manchester United team had just lost at Southampton because they were wearing the wrong colour shirts!

This chapter has highlighted things that matter greatly in the search for UFO answers and only the Maureen Puddy case contains a suggestion of anything alien. For ufology to solve the big mysteries it has to take on board the complexities of all cases. It isn't easy to figure out what all of this might mean.

So What?

We have too many answers chasing too little evidence. As in politics, more people claim they have the answers than deliver progress. One rule I stick to is to distrust any piece of media promising 'the' (as in singular) answer to the UFO mystery. In a much longer book than this John B Alexander made a great point: 'One of the most dangerous positions one can take in an area as complex as UFOs is to be dogmatic about convictions. This is a fatal error and one made by many UFO researchers who have written on this topic.'[1]

There is progress being made in research. More importantly, there is a mystery here substantial enough to excite anyone with courage, a brain and determination. And ufology is big. Big in its ideas, big in the range of answers on offer and big enough to reach the population of the whole planet. Someday, maybe, it will do more than that. It could provide us with answers about who we are and what we are doing here. It almost certainly will provide us with some answers about the world around us.

So what does it all mean? In the absence of proof, you can draw your own conclusions. In the search for conclusions to this book we can do little better than line up the usual suspects and consider their claims.

Space is the Place

Many people believe that UFOs come from space, although the exact place(s) in space are still argued over. Some of the evidence doesn't pass muster in any scientific sense. Religious groups, like the Aetherius Society, who claim contact with inhabitants of our own solar system living in higher spiritual realms, are certainly out on their own. Scientifically, their claims are possible only in terms of discoveries as yet unmade. They have a crude logic but also a raft of detractors within both ufology and science. There's also the uncomfortable fact for some groups that events they have predicted (up to and including the end of the world) have failed to materialise. On the other hand, there are cases of channelled contacts, telepathic messages and the bizarre array of alien forms who arrive with messages and/or a purpose to their mission. Hard evidence or not, there is a pattern to many UFO reports that indicates some intelligence trying to reach us.

It is notable that, in case after case, the witnesses don't seem to question the extraterrestrial origins of the beings they meet. Antonio Villas Boas made love to a being who was certainly humanoid and indicated that she would give birth to his child in space. Maureen Puddy instinctively knew she was dealing with a spaceman.

In terms of the harder evidence there are claims about planets of origin. Some, like those of the many abductees in the last two decades who seem to meet grey aliens, have a consistency. Some claims, based on everything from information gathered at hypnotic regression sessions to alleged leaks of classified information, pinpoint the origins. For the last few decades, Zeta Reticuli has become a particular favourite. The hard evidence for this eludes us. The history of 'definitive proof' in this area is the history of evidence that failed to appear or disappointed when it did. A major reason for the predominance of Zeta Reticuli as a favoured origin for aliens is a 'star map' drawn by Betty Hill following the UFO event she experienced with her husband Barney in 1961. In 1968, Marjorie Fish, a school teacher who was intrigued by the star map as it appeared in *The Interrupted Journey*, John G Fuller's account of the Hill's alleged abduction, interpreted the alignment of stars on the map (which purportedly showed travel routes used by the aliens) to indicate they had come from Zeta Reticuli. Fish's calculations were disputed over 20 years later when the European Hipparcos mission (which collected precise astronomical data) proved that some of the stars identified by Marjorie Fish were actually much further from Earth than had been thought in the mid-1960s.

The Wow! signal has defied all attempts to explain it definitively and therefore remains the best piece of evidence for intelligent extraterrestrial life. There are gaps in all the arguments and the evidence supporting these arguments. The strongest argument against the Extra-Terrestrial Hypothesis (ETH) is the consistent lack of proof and the consistent disintegration of the cases that offered the proof. But claims and mysteries remain and some complex cases, like the incidents at Bentwaters and Tehran, with their suggestion of objects that appear to possess physical capabilities and intelligence or intelligent control beyond the level of our own, do present a challenge to consistent scepticism. It doesn't help the pro-ET area of ufology that so much content is devoted to cases long since explained but still presented as if they are genuine mysteries. It is also problematical that some cases explained in detail rely on the same scenarios (sometimes up to and including the same grey aliens) as many of those now discredited. Indeed, it's for this reason that 'Britain's Roswell' aka the Rendlesham Forest incident is excluded from this short book. Those claiming the case as an interaction between an alien spacecraft and US forces on British soil have a vivid story, a declassified US Air Force document from a reputable source (Charles Halt who was then a Lieutenant-Colonel), and eyewitness testimony in a handful of books and films. Like Roswell there is dispute about the actual landing site (though in this case we're talking a difference of many yards, not miles) and also significant differences in what actually occurred. All of this took place very close to a major military base. So 'Britain's Roswell' is a phrase that makes sense to sceptics and proponents of the Extra-Terrestrial Hypothesis, but each side of that debate brings to mind a different thing when they hear the phrase.

The God Slot

Ufology is a religion. More accurately, it encompasses aspects of organised religious belief. That is fact. From organisations registered as religious charities to ceremonies that see massed Aetherians chanting on holy mountains like Holdstone Down in Devon, there are religions based around UFO belief. Pragmatic and cynical demolitions of such groups exist but most of these groups contain self-aware people, content with their lot. Marshall Applewhite and his Heaven's Gate suicide commandos, who committed communal hara-kiri as the Hale-Bopp comet passed the Earth, are very much the exception in this company.

More debatable is the notion that ufology and the UFO community are a pseudo-religious movement. Some outsiders, especially those who make the study of society their business, say yes. But they'd be eaten alive, metaphorically speaking, by the people they are describing in these claims. Whatever the views of those inside the UFO community, it's true that the issues and ideas that draw some of them in can act as a surrogate religion. Anecdotally the evidence is there, in the way some pieces of evidence are treated as important relics and the way that faith upholds some stories in the absence of proof. Many stories, for example some of those about crash retrievals, are completely items of faith. The occurrence of miracle cures for AIDS and colour-blindness are believed by some in the UFO community, but are unsupported by medical science. As we saw in the last chapter, some of the roots of the crash retrieval stories may be government disinformation. John A Saliba wrote a paper outlining the fact that UFO stories present a range of classic religious ideas: mystery; transcendence; the existence of supernatural entities; images of perfection; ideas of salvation; a world-view that appeals; and an explanation of spirituality.[2] Look through the stories in this book and these themes will emerge. Directly in the case of the vision given to Ezekiel. Indirectly, in the contact between Maureen Puddy and her spaceman who clearly had the power to remain invisible to some whilst revealing himself to her.

The existence of the supernatural and, by implication, another realm that would appear miraculous and perfect to many on Earth, is implicit in the stories of beings who can float humans through walls and windows into waiting spacecraft. Belief in such stories can bind people to UFO groups. To say ufology is a religion or pseudo-religion is probably an overstatement but belief in UFOs does address needs that may also be met by religious teachings. UFO belief continues to rise in a period that has seen the major established religions fragmenting. In an era that has seen a satirical church devoted to the Partridge Family[3] and the serious discussion of religious experiences involving the spirit of Elvis Presley,[4] the notion of the space brothers acting as guardian angels isn't so unusual. The twenty-first century has seen an increasing interest in religion from a secular angle, both in the growing popularity of movements like Humanism and in books like Kevin Nelson's *The God Impulse* (2011) and Reza Aslan's *God: A Human History of Religion* (2017) which map out a pragmatic argument about humanity's propensity to turn its own lived experience into meaningful theology. The extent to which the UFO community sharing their beliefs and needing to be close to experiencers, like abductees, is actually religious behaviour in the widest sense is debatable. It's a guarantee that most of that debate would be very lively.

The Appliance of Science

Ufology is not a recognised science. Not a physical science and not a social science. Many investigators in groups and their counterparts in professional science are openly hostile to each other. Still, UFOs have provided some small scientific breakthroughs. The earthlight theories go hand in hand with research in tectonics and geology. The ideas about electromagnetic hypersensitivity promoted by a handful of researchers are also founded on observation. The groundbreaking work of Michael Persinger and Albert Budden didn't prove the causes of abduction beyond reasonable doubt, but it did establish a body of evidence a few others have continued to explore. In occasional instances – like the work of Project Hessdalen – ufology and science are one.

It won't please many directly within the UFO community but it may well be that some of the key scientific breakthroughs that solve UFO mysteries will come from unrelated studies. It has been realised for years that abduction cases resemble stories of kidnap by fairies in the past. There are also other puzzling events, like Alfred Burtoo's encounter, that produce sincere witnesses and stories that seem to be part logic and part nonsense. In a word, dreamlike. This notion is further supported by the controversial study carried out by Lawson and McCall that at least hinted that we may all be able to produce abduction experiences, even when others know for sure that no abduction took place. All of which suggests that work in psychology and sociology that is already looking at UFO-related experiences will continue to reveal truths about how our minds work and how our experiences influence each other. Some of the most significant breakthroughs in the study of UFOs in the last few decades have come from social science. They haven't always impressed those involved in UFO investigation.

Ironically, science and ufology are furthest apart where their interests are most common – in the search for proof of extraterrestrial life. Consider the roles of those employed in SETI-related work and ufologists and this makes sense. Each side often considers the other engaged in pointless work with no reason to exist. To many ufologists the radio telescopes trained into infinity are a waste of money when they believe the evidence and eyewitness accounts from Rendlesham Forest, Roswell and the rest might deliver what SETI wants to know. There is a certain ironic comedy in this situation if only because the other common ground that ufology and SETI share is a history of throwing huge efforts into searches that produce interesting but marginally useful results. The big prize continues to elude both sides. In one crude sense they are many miles apart in an earthly sense but very close in a cosmic sense. Edward Ashpole's *The UFO Phenomena* (1995) makes an erudite argument for the sense in looking for UFOs within our own solar system but not, necessarily, on our planet. His point is that we know from our own experience that space probes are relatively safe from harm, until they risk entering the gravity and atmosphere of a planet. He reasons like other scientists that probes are more efficient than sending living entities to investigate distant space, not least because they could be programmed to remain dormant until something interesting – like detecting a radio signal – wakes them up.

But this prolonged animosity is not the whole picture. There are a handful of researchers on either side who move easily between both areas and there are also periodic attempts to combine forces. In the case of abductions, the Massachusetts Institute of Technology hosted a high-powered symposium in 1992, recorded in CDB Bryan's book *Close Encounters of the Fourth Kind*, that brought the leaders of abduction research face to face with social scientists and a smattering of other interested parties. There were some volatile eruptions, especially when those in the UFO community began discussing the research methods that allowed them to estimate the number of people being abducted. But there was progress and understanding too.

There is no doubt that social and physical science can help those involved in UFO research to make progress. There is also no doubt that evidence gathered in the course of UFO investigation could help researchers in the sciences. In the physical sciences, phenomena like ball lightning remain mysterious. In social science, conditions like Fantasy Prone Personality remain to be quantified and fully understood. The hard evidence collected by ufologists may be useful to these sciences. For UFO researchers genuinely interested in solving cases, the occasional chance to work directly with professionals is a bonus. It is probably, also, the best chance for the subject of ufology to gain widespread credibility and respect.

And, of course, science is still pouring efforts into making sense of the Wow! signal, or collecting another one.

You Don't Have to Be Mad to Work Here...

Linked to the social scientific investigations into UFOs is the central question of whether any of the experiences described can be explained by psychology. We're long past dismissing the whole thing as madness but we cannot say that all experiences described are genuine, physical encounters. The major obstacles to definite statements either way include a lack of evidence on the clinical side and a confusion of evidence in the area of UFO experience. One thing we can safely say is that no one trigger experience brings about UFO events. In some cases, people simply misidentify things. Meteors, the planet Venus and aircraft regularly prompt UFO reports. The people making the reports generally make them in all sincerity but UFO researchers are often wise to the patterns of certain types of misidentifications. Jenny Randles' UFO Study (1986) included observations on how things as mundane as a swarm of insects can produce a predictable kind of sighting report.

The misidentifications go to the heart of the business. The Kenneth Arnold sighting that started the modern era of ufology may well have been a misidentification. At one end of the psychological spectrum we have simple slips of identification. At the other we are dealing with full-blown altered states of consciousness serious enough to take over someone's life. There are contradictions in the messages channelled and gathered directly from aliens. This proves, at least, that some channelled contacts and direct encounters with aliens are not what they seem. This leaves us with the problem of explaining the causes of such messages. It is likely that the future will show developments in understanding psychological conditions that will advance both clinical practice and the understanding of UFO experience.

The major reason we cannot reach conclusions at the present time is because we lack definitive diagnostic tools for conditions like Fantasy Prone Personality Disorder (FPP). Most of the conditions that might explain UFO encounters are either the stuff of peer-reviewed journals or established within areas of clinical practice that don't generally bring the practitioners into contact with UFO investigation. FPP is firmly within the former category. It is not presented in the DSM (Diagnostic and Statistical Manual) series of books, which are the definitive diagnostic tools for clinical practice. FPP is an item of debate and discussion amongst some therapists and it is often the case that a psychological condition debated and discussed in this way falls short of a recognised disorder, but does find its symptoms catalogued and used to aid the understanding of patients or clients.

Some established conditions may provide insights into existing UFO cases. Lynn Picknett's *Mammoth Book of UFOs* considered a visual disorder, Charles Bonnet Syndrome (CBS), as a potential basis for explaining sightings of beings or craft that were not seen by others. As explanations for UFO events go, CBS is an outsider at present. It is a visual disorder that affects the elderly and is only found in patients who have suffered damage to their sight. You could not easily explain, for example, Maureen Puddy's encounter in this way. She was 37 at the time of the main event described in this book. However, CBS does establish beyond doubt that sane elderly people with damaged eyes can see things that are not there and they often perceive the hallucinatory visions for what they are.[5] It also establishes that the relationship between the nervous system controlling our eyes and the centres of the brain concerned with the creation of meaning is complex. CBS won't explain Maureen Puddy, Alfred Burtoo, Joe Simonton or the many others whose encounters with craft and occupants present vivid experiences for the witness and little or no corroboration from others. However, it may be that in the clinical literature on a range of conditions, such as epilepsy, there are nuggets of information that will eventually lead to such an understanding.

Ongoing investigations need evidence. Psychological investigations usually start with a hard core of cases in which a compelling nucleus of evidence presents researchers with the basis for diagnosing a condition. At this point, understanding the condition depends on exploring a larger body of evidence. For this reason, it is likely that many of the major breakthroughs in UFO investigation will come from neuroscience and psychology. This will happen because there will be time and money for research. In some cases, those doing the research will see the sense, as Dr Michael Persinger did, in using UFO accounts and claimants as the basis for investigation. Such work will happen but it is impossible to predict where it will take us.

Finally in this section, there is an uncomfortable but unavoidable issue which is that a few in the UFO community have made contributions to the stories that appear in UFO material but their contributions may well have gone hand in hand with their own mental health conditions. The grim details need not detain us here but if you wanted to explore an extreme and very uncomfortable case for ufology then going online to investigate the life of Paul Bennewitz (1927-2003) might prove an instructive example. It's a very extreme example but instructive in the sense that people like Bennewitz are drawn to involvement in the UFO community at which point they influence others but remain rife for manipulation themselves.

The Infotainment Scam

It's likely you bought this book or borrowed it from someone who paid for it. This book exists because a commercial publisher saw the financial sense in adding a title to their range of punchy and informative works. UFOs may be a lifelong interest of mine but I have also been a professional creative throughout my adult life. So, part of my working life involves taking what I know and shaping it for a market. With the subject of UFOs I have a problem. The books that sell millions of copies are the ones that claim that the aliens are here and coming for us all, and I was never likely to write a book like that. If I self-published or went through a small, anonymous publishing house, then the chances of the book getting decent sales would be very limited. Oldcastle Books on the other hand have a house style, a track record of no-nonsense writing, a sales team and a presence in the major bookshop chains.

This is the most pertinent example I can show you of the value of UFO information, because in this case the market included you, personally. The information in this book had a value before the book existed. Its value to the publisher is proven because they invested money and commissioned the title. They are not short of subject ideas or potential writers. Their survival depends on making profitable decisions from these options. If I wanted to make money as an author, I was always going to be better off putting my work with a company who can organise distribution of the finished product, have a track record of selling books and an established style and market.

Work outwards from this example and we find a key point of ufology. Outside of direct research into events, or more accurately, reports of events, the information we gather is usually produced commercially. Academic research may be one exception to this rule but most of the academic papers ufologists own on their subject are compiled in commercially available books and the work is influenced by priorities in academic departments, which means it is also linked to issues like funding. This is one reason why areas like psychology and sociology are the leaders in publishing academic papers with some relevance to ufology.

The market for UFO information has consistently liked work claiming aliens are real, governments cover things up and the proof of both of these claims is in the book/magazine/website/ video you are about to consume. This is a situation that has cast UFOs and aliens as a potent market force. More importantly, this situation has ensured an abundance of pro-extraterrestrial material at the expense of other views. In cases like Roswell, the various views exist in print, partly, because the name recognition alone is a market winner.

One obvious effect of this situation is to ensure that the definitive truth will, almost certainly, never be known. There are precedents for this. For example, the death of Marilyn Monroe and the identity of Jack the Ripper remain shrouded in mystery partly because the subjects have shown themselves to be consistently lucrative as media source material. Much material sells on the basis of offering a revision or new angle. Like the Roswell case, the only realistic hope of an end to the speculation is the emergence of a new piece of definitive evidence. This is an unlikely prospect in all these cases.

The market-led nature of UFO information often harms the search for the truth. I made that particular observation in an earlier version of this book in 2002. Since then the internet's influence has become pervasive. Lots of material online looks and sounds impressive but the video and audio trickery available to many people now makes it hard to trust much of this material. The tonnage of new content being added every day and the fact some of it goes viral means one aspect of ufology now involves an online community continually feeding itself and attracting new members. Algorithms typically direct users to more and more content, not deeper and deeper knowledge. Faced with the content and isolated from the content creators we might all struggle to separate mystery from hoax and serious intent from serious fun. This situation is at the extreme end of a dynamic in which an adherence to the truth and an awareness of market needs make an uneasy alliance. Remember the point that Timothy Good made in his *Guardian* interview promoting the massive-selling *Beyond Top Secret*: 'If ever there's a subject that needs rescuing from its supporters, it's this one.' Indeed the digital age has proven Good's point in one succinct way because the ease of researching historic information has allowed the more sceptical researchers to find numerous cases in which stories central to UFO mysteries can be sourced to a single and very questionable appearance in a book or article, and simply grew in the telling. A blatant example of this is the core of the *Skeptoid* episode 'The Vanishing Village of Angikuni Lake' (2013), which shows how an improbable tale, written vividly by early UFO author Frank Edwards, gradually grew to describe a mass disappearance. *Skeptoid* quickly established there was no hard evidence in Canadian police records, or the contemporary press to support any of the claims.[6]

During the mid- to late-1990s, it was widely claimed that surfing on the internet for UFOs, the paranormal and conspiracy theories was second only to surfing for sex sites in popularity. When the Blair government in the UK began dealing with the requests that poured in after they passed a Freedom of Information Act they were amazed at the level of demand for UFO information. That outcome didn't surprise people in the UFO community. It led, eventually and predictably, to such non-fiction books as *The UFO Files: The Inside Story of Real-life Sightings* (2012) and –covering paranormal cases in general – *Britain's X-Traordinary Files* (2014). Both worthy reads, but more proof that there is money to be made in the subject and that infotainment remains a big part of why the UFO mystery is so well known.

The point, I hope, is made. UFO and alien ideas are widely understood, potentially profitable and adaptable over time. Incredibly so if we take a second to think how much popular fiction – like the *Alien* series of films – is based on some of the ideas that underpin ufology. This matters because one theme running through nearly all of the cases outlined in this book is that many people encountering aliens, apparently for real, are also confronted with a reality that makes a general kind of sense. In most cases, aliens appear to be ahead of us in terms of technology and also have an awareness of the problems faced by the human race. This picture is, more or less, true even when the encounters don't involve space aliens. The American airship wave of the late nineteenth century included James Hooten's meeting with an airship captain who used 'compressed air and aeroplanes'.

The whole airship wave was, almost certainly, fiction. However, the audience didn't always think so. The 1950s contactees similarly had a large following, many of whom believed every word. Today aliens impart a different wisdom or carry out interbreeding programmes, possibly to benefit mankind and themselves, or they simply appear to exhibit levels of technical brilliance to which our own engineers aspire. In every generation we appear to have some kind of archetypal story of encountering other realms and realities. In the digital age we have access to information on an unprecedented scale and almost everything can be accessed, literally, at the speed of light. 5G brings our household devices into the conversation and moves our whole lives further into cyberspace, adding another layer of complexity.

Within this growing web of possibilities, the audiences have countless choices of what to read, watch, surf and experience. More than ever now, we choose what we become. Where the UFO experience is concerned the big question remains how much we choose it and how much it finds us.

So Farewell Then

This factual book can't offer closure on its subject. There are few formulae or rules to memorise that will make you an instant UFO expert. The more you look for answers, the more you find questions. Ufology is puzzling, downright frightening, frequently hilarious and always capable of surprising you. The title of this book didn't lie, finding the truth is often a battle.

I've learned some survival strategies over the years. Confronted with someone who claims to have hard evidence in *the* definitive case I'm no longer impressed. I've spared you the main reason for that last statement – ufology's history of presenting a major case that claimed to be able to change everything, before it crumbled to dust. Feel free to Google the MJ-12 documents, Manhattan Transfer Abduction, 1989 Kahari UFO crash or the Alien Autopsy Movie to explore that angle. When I hear another variation of the cover-up and conspiracy stories I am, once again, inclined to cynicism. But it is an odd kind of disbelief because I'd love to be wrong. In fact, if some major political figure throws open the hangar doors and shows us all the antimatter reactor that arrived with the alien craft, I'd be happy to have got it wrong over the years.

But I've also learned to love ufology in all its ramshackle glory because its ability to teach us about ourselves is incredible. It highlights the fringes of science, it may well be teaching us more than we realise about ourselves and our behaviour both as individuals and collectively, and, in its focus on our biggest concerns about who we are and where we fit into the great scheme of things, it helps us to ask some really fundamental questions. It's also an incredible if unofficial insight into our history. Seriously, the benefit of reading books from decades ago is vastly undervalued. You realise quickly that some of the realities of this subject have changed massively over time. The aliens sighted in the 1960s, for example, were a bizarre and often surreal menagerie of improbable life forms. The greys were there but their rise to complete dominance started around the time they appeared in the closing scenes of *Close Encounters of the Third Kind* (1977). Similarly, Roswell – for all its prominence at the close of the last century – is conspicuously absent from most UFO literature of the preceding decades. Stories in *The National Enquirer* in the late 1970s, followed by the massive-selling *The Roswell Incident* (1980), provided watershed moments for that particular case. One advantage of the internet age is the republication and mass second-hand availability of everything from the bizarre contactee tales of the 1950s to the complete books devoted to cases now. The way these messages are presented, what we can learn about the expectations of the audience for them and the changing nature of the accounts all present insights. Crudely, through looking at the aspect of this subject that an academic would call the 'discourse', we can learn a lot that helps us analyse the things presented in the present day. The internet offers more sites and more information than any one dedicated amateur can hope to keep up with, after which you can relax by binge-watching a series of *Hangar 1 – The UFO Files*. None of this avalanche of material shows any sign of slowing down. If you want to talk about your stress levels in trying to keep up, you could always buy a ticket to a conference, like those run by *Outer Limits Magazine* in the UK.

As I wrote at the start, this is a subject that offers people a lifetime of lateral thinking. It is also a subject that allows an amateur to make a major contribution. So, if you're lacking that sense of closure and still wondering what happened at Roswell or over the Bass Strait, there is one solution open to you: get yourself involved, preferably through attending a conference or some other means of actually meeting the people who take an active part in the whole business. If you read, or surf, it really helps in the long run to investigate the stuff that most disagrees with your own viewpoint. The truth most certainly is out there – but it'll be a hell of a lot easier to find it if we use some critical thinking before we start asking the questions.

Notes

Books, websites, articles and live presentations pillage the same cases. In the best-known cases, like Roswell, the sources quoted here are chosen because, in the opinion of this author, they present an accessible and authoritative way to follow up the introduction provided by this book.

Introduction

1. Leonard Stringfield, 'The Search For Proof In A Squirrel's Cage' in Hilary Evans & John Spencer, *UFOs 1947-1987: The 40-Year Search For An Explanation*, Fortean Times 1987, pp 145-155.

2. Joseph Allen Hynek (1910-1986) arguably did for UFO studies what Elvis Presley did for popular music. A scientist with a PhD in Astrophysics who first became involved in the subject via official investigations, Hynek was initially of a sceptical persuasion and – briefly – became a figure of some vilification when the press and public were unconvinced by his assertion that a series of UFO events in Michigan may have been caused by swamp gas igniting. It was unfair on a man whose allegiance to the 'Hippocratic oath of a scientist'

meant he pursued truth above anything else. This led Hynek to write *The UFO Experience* (1972), *The Edge of Reality* (1975) and *The Hynek UFO Report* (1977). In these books, he stated his view that UFOs represented a compelling and substantial challenge to established scientific thought. In an area troubled, then as now, with sensational writing and sketchy research, Hynek's work allowed the subject to raise itself towards respectability.

3. David Clarke and Andy Roberts, *Flying Saucerers: A Social History of Ufology*, Alternative Albion Press 2007. Long out of print and frequently eye-wateringly expensive when offered second-hand. The Armchair Ufologist is long discontinued as Roberts' UFO career has been on voluntary hold for years. I can't improve on Jenny Randles' description of Roberts in *The Little Giant UFO Encyclopaedia* as an 'irreverent wit and rational UFO investigator.' David Clarke – by contrast – continues a significant involvement in the field alongside his professional career as an academic.

4. James Oberg, 'The Failure of the "Science" of Ufology', *New Scientist* 11 Oct 1979, pp 102-106.

5. Neil Nixon, 'Ufology – A Qualified Success?', ed. Steve Moore, *Fortean Studies 5*, John Brown Publications 1998, pp 96-121.

6. Brian Dunning. Award-winning host of the *Skeptoid* podcast, sceptically inclined author on the paranormal and pop culture and a tireless advocate of detailed investigation and – where possible – peer-reviewed research being applied to all of the above.

7. Erich Von Daniken, *Chariots of the Gods*, Souvenir 1969. Von Daniken remains the bestselling and best known of a

slew of authors who have popularised this branch of UFO investigation.

8. Josef F Blumrich, *The Spaceships of Ezekiel*, Corgi 1974.

9. Carl Sagan (1934-1997) was another hugely influential thinker and investigator in the field.

10. The classics of which include John Michell's *The Flying Saucer Vision* (1967) and *The View Over Atlantis* (1969), and Ronald Story's *The Space Gods Revealed* (1976).

11. Robert Temple, *The Sirius Mystery*, Futura 1976.

12. David Barclay, *Aliens: The Final Answer?*, Blandford 1995.

13. A short and learned demolition was available here at the time this book was written: https://www.smithsonianmag.com/science-nature/the-idiocy-fabrications-and-lies-of-ancient-aliens-86294030/.

14. An insightful discussion, including a discussion on Charlemagne's alleged banning of UFOs was available on Jason Colavito's blog at the time of writing: http://www.jasoncolavito.com/blog/did-charlemagne-legislate-against-ufos.

15. Dr David Clarke, 'Once Upon A Time In The West', Chapter 1 of *The UFOs That Never Were*, Jenny Randles, Andy Roberts & David Clarke, London House 2000, pp 15-32.

16. Jaques Vallée, *Challenge To Science*, Neville Spearman 1966 and *Passport to Magonia*, Neville Spearman 1970.

17. Jaques Vallée, *Passport to Magonia*, Neville Spearman 1970, pp 23-25.

18. Margaret Sachs (ed) *The UFO Encylopedia* Corgi Books 1980, pp 87-88.

19. David Barclay, *Aliens: The Final Answer?*, Blandford 1995, pp 82-83.

20. James Easton, 'Flight of Fantasy,' *Fortean Times* 137.

21. Hilary Evans & Dennis Stacy, *The UFO Mystery*, John Brown Publications 1998, pp 28-29.

22. Tim Shawcross, *The Roswell File*, Bloomsbury 1997, pp 142-144.

23. Ed David C Knight, *UFOs: A Pictorial History from Antiquity to the Present*, McGraw-Hill 1979, pp 31-33.

24. Jenny Randles & Peter Hough, *The Complete Book of UFOs*, Piatkus 1994, pp 122-129.

25. A potted overview of the case appears in Alan Baker, *The Encyclopaedia of Alien Encounters*, Virgin 1999, pp 19-20.

26. Curtis Peebles, *Watch The Skies: A Chronicle of The Flying Saucer Myth*, Berkley Books 1995, pp 198-199.

27. Jenny Randles, *Aliens: The Real Story*, Hale 1993, p 29.

28. Robert Sheaffer, *Bad UFOs: Critical Thinking about UFO Claims*, Self-Published, 2016, pp 126-127.

The Evidence for Alien Invaders

(Firstly, a note about the notes in this section. The links in some cases are directly to the most sceptical views of those cases. This isn't a case of favouring one side of the argument over the other so much as simple common sense. Where the positive arguments about these cases are concerned there is no shortage of material to be found after a few seconds of an internet search.)

1. Timothy Good, *Above Top Secret*, Sidgwick and Jackson 1987, and *Beyond Top Secret*, Sidgwick and Jackson 1996.

2. Relax, it didn't.

3. Timothy Good, *Alien Liaison*, Arrow 1992. Lazar's claims

are a major part of this book, the hard evidence of the payslip is on page 178.

4. John E Mack, *Abduction*, Simon and Schuster 1994, pp 41-43.

5. Jenny Randles, *UFO Retrievals*, Blandford 1995, pp 18-25. The whole book comes highly recommended as the best and most accessible collection of UFO crash stories.

6. Link to short article discussing claims of the Trinidade pictures being hoaxed: http://forgetomori.com/2010/ufos/trindade-island-case-photographer-admits-hoax/.

7. *Skeptical Inquirer* article suggesting Paul Trent's photos are hoaxed: https://skepticalinquirer.org/2015/01/the_trent_ufo_photosbest_of_all_timefinally_busted/. And a 2013 report covering the same ground in more detail: http://www.ipaco.fr/ReportMcMinnville.pdf.

8. Jenny Randles, Andy Roberts & David Clarke, *The UFOs That Never Were*, London House 2000, pp 196-197.

9. Jenny Randles, *Something in the Air*, Hale 1998, pp 124-131.

10. Report from the *Independent* on the Underwood/Fravor case: https://www.independent.co.uk/news/science/ufo-tic-tac-flying-saucer-chad-underwoord-dave-fravor-a9254671.html.

11. *Skeptoid*, 'The Pentagon's UFO Hunt', 1 May 2018: https://skeptoid.com/episodes/4621.

12. 'Alien' Caught on Camera in La Junta, YouTube 12 June 2019: https://www.youtube.com/watch?v=gFy6LVlNc6c.

13. The Iron Skeptic's scathing analysis of the Falcon Lake case: http://www.theironskeptic.com/articles/michalak/michalak.htm.

14. The more scientifically minded UFO books love the

Trans-en-Provence case. Try Edward Ashpole, *The UFO Phenomena*, Headline 1995, pp 116-125, or Peter A Sturrock, *The UFO Enigma: A New Review of the Physical Evidence*, Aspect 2000, pp. 257-297.

15. CDB Bryan, *Close Encounters of the Fourth Kind*, Weidenfeld and Nicholson 1995, pp 19-20 & 26.

16. Jenny Randles, *Aliens: The Real Story*, Hale 1993, pp 61-63.

17. Malcolm Robinson, *The Dechmont Woods UFO Incident: An Ordinary Day, An Extraordinary Event*, Self-Published, 2019, pp 118-210.

18. Jenny Randles, *UFO Reality: A Critical Look at the Physical Evidence*, Hale 1983, pp 68-69, 70-71 & 88.

The UFO Community

1. Robert Durant, 'Public Opinion Polls and UFOs' in Hilary Evans & Dennis Stacy, *The UFO Mystery*, John Brown Publications 1998, pp 338-352.

2. Full report online here: https://news.gallup.com/poll/266441/americans-skeptical-ufos-say-government-knows.aspx.

3. Ken Phillips, 'The Anamnesis Report', *Proceedings of the 6th International UFO Congress*, pp 59-63.

4. Ken Phillips, 'The Psycho-Sociology of UFOs', ed. David Barclay, *UFOs: The Final Answer*, Blandford 1993, pp 40-64.

5. Judy O Parnell & R Leo Sprinkle, 'Personality Characteristics of Persons Who Claim UFO Experiences', *Journal of UFO Studies* Vol 2, CUFOS 1990, pp 54-58.

6. SC Wilson & TX Barber, 'Vivid Fantasy and Hallucinatory

Abilities in the Life Histories of Excellent Hypnotic Subjects', ed. E Klinger, *Imagery Volume 2: Concepts, Results, and Applications*, Plenum Press 1981.

7. Robert E Bartholomew & George S Howard, *UFOs and Alien Contact*, Prometheus Books 1998, pp 248-273.

8. Chris Rutkowski, *Abductions and Aliens*, Fusion Press 2000, pp 162-163.

9. A short write up of Knighton's case is still available online here: http://www.ufosoveramerica.com/html/ufo_stories.html.

10. Steven P Resta, 'The Relationship of Anomie and Externality to Strength of Belief in Unidentified Flying Objects', Masters dissertation, Loyola College Graduate School, Baltimore 1975.

11. Martin Kottmeyer, 'UFOlogy as an Evolving System of Paranoia', *UFO Conspiracy Theories* Vol 7, No. 3 May/June 1992, pp 28-35.

12. David Morris, *The Masks of Lucifer*, Batsford 1990.

13. Vance Packard, *The Status Seekers*, Penguin 1963, p 44.

14. Jodi Dean, *Aliens in America: Conspiracy Cultures from Outerspace to Cyberspace*, Cornell University Press 1998, p 60.

15. Dr R Leo Sprinkle, *Soul Samples*, Granite Publishing 1999.

16. Devereux and Budden have both written a series of books charting stages of their evolving studies. The most recent titles are, therefore, recommended. Probably the most accessible view into a radical take on many UFO events is Albert Budden's *Electric UFOs*, Blandford 1998.

17. Edward Ashpole, *The UFO Phenomena*, Headline 1995, p 214.

18. A brief overview of Noyes' background, views and unique

character is here: http://www.openminds.tv/mod-official-in-uk-files-760/11506.

19. Jenny Randles, *Star Children*, Hale 1994, pp 161-182 provides an insightful discussion on UFO states of mind.

20. See Pocket Essential *Conspiracy Theories*, p 55. The book also contains material on UFO conspiracies. RE Bartholomew's 'The Social Psychology of "Epidemic" Koro' appears in *International Journal of Social Psychiatry* 40, no.1 1994, pp 46-60.

Amazing Tales

1. The Fund for UFO Research (FUFOR) was active from 1979-2011, spending around $700,000 during that period with the aim of increasing scholarly research into UFOs and the Extra-Terrestrial Hypothesis and to gain the release of secret documents.

2. Karl Pflock, *Roswell: Inconvenient Facts and the Will to Believe*, Prometheus Books 2001.

3. Try Tim Shawcross, *The Roswell File*, Bloomsbury 1997, which is a succinct and engaging consideration of the pro-UFO case – from a time when many still thought the infamous alien autopsy movie to be the real deal.

4. The Bentwaters case is a well-trodden event in UFO history. Jenny Randles, for one, has revisited the case several times. Martin Lawrence Shough's chapter, 'Distant Contact: Radar/Visual Encounter At Bentwaters', eds. John Spencer & Hilary Evans, *Phenomenon*, Futura 1988, lacks some of Randles' recent forays into official documents but does

provide one of the most vivid and authoritative tellings of the tale.

5. Jenny Randles, *Something in the Air*, Robert Hale 1998, pp 95-97.

6. Peter Hough & Moyshe Kalman, *The Truth About Alien Abductions*, Blandford, pp 8-18.

7. As of this writing the interviews conducted by the British researchers form the final section of the Wikipedia page on the Lakenheath-Bentwaters Incident.

8. Philip Klass, *UFOs: The Public Deceived*, Prometheus Books 1983, pp 111-124.

9. Brian Dunning, *Skeptoid* 385: 'The Disappearance of Frederick Valentich', 22 October 2013: https://skeptoid.com/episodes/4385.

10. Jenny Randles, *Something in the Air*, Robert Hale 1998, pp 139-144.

11. Timothy Good, *Beyond Top Secret*, Sidgwick and Jackson 1996, pp 87-93.

12. UFO Insight web page on the Alfred Burtoo incident with video links featuring Timothy Good speaking about it: https://www.ufoinsight.com/bizarre-ordeal-alfred-burtoo-abduction-wasnt/.

13. Abductions form a major part of the UFO literature and anyone interested in catching up on the classic accounts from the 1990s heyday of abductions is recommended to investigate the work of Budd Hopkins and his Intruders Foundation. Hopkins' books, especially *Intruders*, Sphere 1987. Also worth hunting down are: David Jacobs' work, especially *Alien Encounters*, Virgin 1992; Dr John Mack's *Abduction*, Simon and Schuster 1994; and CDB Bryan's *Close Encounters of the Fourth Kind*, Weidenfeld and

Nicholson 1995. The latter is an account by a previously sceptical journalist of his attendance at an abduction conference and subsequent meetings with some of the main players.

14. Adam Frank, *Cosmos and Culture: Talking to Aliens from Outer Space*, 10 July 2012: https://www.npr.org/sections/13.7/2012/07/10/156540615/talking-to-aliens-from-outer-space?t=1580161497158.

15. Brian Dunning, *Skeptoid* 342: 'Was the Wow! Signal Alien?', 25 December 2012: https://skeptoid.com/episodes/4342.

Cautionary Tales

1. The *Skeptical Inquirer*'s take on the Phoenix lights: https://skepticalinquirer.org/2016/11/the_phoenix_lights_become_an_incident/.

2. Story from William L Moore's article 'The Crown of St Stephen and Crashed UFOs: The Oldest UFO Disinformation Case on Record', *Far Out* Vol 1, No. 1 July 1992, pp 32-33. For a thorough round-up of crash retrieval stories see Jenny Randles, *UFO Retrievals: The Recovery of Alien Spacecraft*, Blandford 1995.

3. Albert Budden, *Electric UFOs*, Blandford 1998, pp 262-8.

4. CDB Bryan, *Close Encounters of the Fourth Kind*, Weidenfeld and Nicholson 1995, pp 70-75 covers the presentation of Puddy's case at the international symposium held at the Massachusetts Institute of Technology in 1992.

5. Useful overview of the Maureen Puddy case online here: https://hauntedauckland.com/site/maureen-puddy-case/.

6. Odd-Gunnar Roed, 'Project Hessdalen', *6th International*

Congress, UFOs: The Global View, BUFORA/IUN 1991, pp 55-58.

7.   David Clarke & Andy Roberts, *Phantoms of the Sky*, Robert Hale 1990, pp 145-146.

8.   Jim Schnabel, *Dark White*, Hamish Hamilton 1994, pp 125-126.

9.   Robin Ramsay, *Conspiracy Theories*, Pocket Essentials 2000, p 78 places this event in a worldwide operation to test mind-control technology.

So What?

1.   John B Alexander, *UFOs: Myths, Conspiracies and Realities*, Thomas Dunne Books 2011, p 222.

2.   John A Saliba, 'Religious Dimensions of the UFO Abductee Experience', ed. James R Lewis, *The Gods Have Landed*, State University Of New York Press 1995, pp 15-64.

3.   The Partridge Family Temple, headed by the Most Reverend Point Me In The Direction Of Albuquerque Partridge.

4.   A useful primer on the subject being Ted Harrison's *The Death and Resurrection of Elvis Presley*, Reaktion Books 2016.

5.   National Health Service page on Charles Bonnet Syndrome: https://www.nhs.uk/conditions/charles-bonnet-syndrome/.

6.   *Skeptoid* 371, 'The Vanishing Village of Angikuni Lake', 16 July 2013, https://skeptoid.com/episodes/4371.

Surfers' Paradise

First, a note – there is no shortage of books, television documentaries and other material on this subject. To avoid a lengthy list of stuff that is very easy to find I've left those items off this page and stuck to material that is online or forward-looking (i.e. magazines still publishing or conferences and podcasts still producing) so the main point of this list is to give you resources that will keep you up to date, and to keep that list very brief. Frankly, it isn't that hard to find enough material to bury yourself in this subject.

Websites

Abovetopsecret.com – http://www.abovetopsecret.com/index.php
Seething cauldron of conspiracy which includes a massive section on aliens, UFOs and related conspiracy material. Extensive archive includes documents and video/still images.

Badufos Blogspot- https://badufos.blogspot.com/
One of two sites operated by arch-sceptic Robert Shaeffer and updated on a regular basis with insights and nuggets of information designed to challenge the sloppier practitioners in

research and those keener on making money than searching for the truth.

Barry Greenwood UFO Archive – http://www.greenwoodufoarchive.com/
One of many personal obsession/public treasure sites online. Barry Greenwood basically trawls obscure and old publications for articles and continually updates the collection, offering scans and simple navigation to bring the original material to a wider public than those creating it years ago ever imagined. As with other homes of obscurities (like many music blogs) this is a two-way street and there's a wants list on display.

BUFORA – https://www.bufora.org.uk/
Aka the British UFO Research Association. It pains me to say it's been a very low-key site for several years. This organisation was the target of much of the scurrilous comment in *The New Ufologist* when it was published but the serious intent of the barbs was to remind everyone life didn't have to be this way. Long established and with an illustrious history, BUFORA has been at the heart of much of the best UFO investigation done in the UK and still serves to train researchers and coordinate understanding on the subject.

CUFOS – http://www.cufos.org/
Aka the Centre for UFO Studies, which continues to follow the example set by the legendary J Allen Hynek (arguably the greatest ufologist of all time). A mine of information, informative articles and a useful one-stop shop for a newbie willing to dive in at the deep end from the very start.

Isaac Koi – https://www.isaackoi.com/
Pretty much what it claims to be, a massive and ever-expanding online archive presided over by a British barrister who practises law but – on this evidence – lives and breathes ufology. Amongst the most useful features is a section allowing you to find an authoritative list of sources citing references in a range of published material which will allow you to explore individual cases and the work of the leading figures in the field. The free researchers starter pack is a bonus too – no pressure but it's clearly there to encourage involvement and critical thinking!

MUFON – https://www.mufon.com/
Aka the Mutual UFO Network; long-established, US-based organisation with a site that passes as slick and well-presented by the standards of the UFO community. Largely pro-extraterrestrial but also broad-minded enough to provide results for most shades of opinion and presented by an organisation that remains substantial and highly active.

Skeptoid.com – https://skeptoid.com/
Site built on a listener-supported podcast which takes on paranormal claims and conspiracy culture with a combination of critical thinking and common sense. In some episodes classic cases (Angikuni Lake) are beaten to a metaphorical pulp; elsewhere (as in the episode on the Wow! signal), you realise there is still much to learn. Award-winning (for the podcast) and latterly expanded to a small empire including video and audio content available for sale.

The National Archives – https://www.nationalarchives.gov.uk/help-with-your-research/research-guides/ufos/

The stuff made available after Freedom of Information legislation opened the floodgates in the UK. Basically a tonnage of information but guidance is made a bit easier with a video introduction by David Clarke (who spent years researching this material and wrote books about what he discovered) and a keyword search option on the site.

Ufo Evidence.com – http://www.ufoevidence.org/

By no means the liveliest or most tended UFO site but worth a look because it's largely a collection of the better known and better evidenced cases submitted to a scientific overview and presented with a pithy journalistic precision that allows the arguments to be followed without too much trouble even when some of the science involved baffles the non-scientist.

Index

●LDCASTLE BOOKS

POSSIBLY THE UK'S SMALLEST
INDEPENDENT PUBLISHING GROUP

Oldcastle Books is an independent publishing company formed in 1985 dedicated to providing an eclectic range of titles with a nod to the popular culture of the day.

Imprints vary from the award winning crime fiction list, NO EXIT PRESS, to lists about the film industry, KAMERA BOOKS & CREATIVE ESSENTIALS. We have dabbled in the classics, with PULP! THE CLASSICS, taken a punt on gambling books with HIGH STAKES, provided in-depth overviews with POCKET ESSENTIALS and covered a wide range in the eponymous OLDCASTLE BOOKS list. Most recently we have welcomed two new digital first sister imprints with THE CRIME & MYSTERY CLUB and VERVE, home to great, original, page-turning fiction.

oldcastlebooks.com

OLDCASTLE BOOKS	KAMERA BOOKS	HIGHSTAKES PUBLISHING
POCKET ESSENTIALS	CREATIVE ESSENTIALS	THE CRIME & MYSTERY CLUB
NO EXIT PRESS	PULP! THE CLASSICS	VERVE BOOKS